REGRESSION MODELS FOR CATEGORICAL AND COUNT DATA

THE SAGE QUANTITATIVE RESEARCH KIT

Regression Models for Categorical and Count Data by *Peter Martin* is the 8th volume in *The SAGE Quantitative Research Kit*. This book can be used together with the other titles in the *Kit* as a comprehensive guide to the process of doing quantitative research, but is equally valuable on its own as a practical introduction to regression analysis with categorical and count data.

Editors of The SAGE Quantitative Research Kit:

Malcolm Williams – *Cardiff University, UK*

Richard D. Wiggins – *UCL Social Research Institute, UK*

D. Betsy McCoach – *University of Connecticut, USA*

Founding editor:

The late W. Paul Vogt – *Illinois State University, USA*

REGRESSION MODELS FOR CATEGORICAL AND COUNT DATA

PETER MARTIN

SAGE

Los Angeles | London | New Delhi
Singapore | Washington DC | Melbourne

THE SAGE QUANTITATIVE RESEARCH KIT

Los Angeles | London | New Delhi
Singapore | Washington DC | Melbourne

SAGE Publications Ltd
1 Oliver's Yard
55 City Road
London EC1Y 1SP

SAGE Publications Inc.
2455 Teller Road
Thousand Oaks, California 91320

SAGE Publications India Pvt Ltd
B 1/I 1 Mohan Cooperative Industrial Area
Mathura Road
New Delhi 110 044

SAGE Publications Asia-Pacific Pte Ltd
3 Church Street
#10-04 Samsung Hub
Singapore 049483

© Peter Martin 2021

This volume published as part of *The SAGE Quantitative Research Kit* (2021), edited by Malcolm Williams, Richard D. Wiggins and D. Betsy McCoach.

Editor: Jai Seaman
Assistant editor: Charlotte Bush
Production editor: Manmeet Kaur Tura
Copyeditor: QuADS Prepress Pvt Ltd
Proofreader: Elaine Leek
Indexer: Cathryn Pritchard
Marketing manager: Susheel Gokarakonda
Cover design: Shaun Mercier
Typeset by: C&M Digitals (P) Ltd, Chennai, India

Library of Congress Control Number: 2021935562

British Library Cataloguing in Publication data

A catalogue record for this book is available from the British Library

ISBN 978-1-5297-6126-9

CONTENTS

LIST OF FIGURES, TABLES AND BOXES

List of figures

List of tables

List of boxes

ABOUT THE AUTHOR

Peter Martin is Lecturer in Applied Statistics at University College London. He has taught statistics to students of sociology, psychology, epidemiology and other disciplines since 2003. One of the joys of being a statistician is that it opens doors to research collaborations with many people in diverse fields. Dr Martin has been involved in investigations in life course research, survey methodology and the analysis of racism. In recent years, his research has focused on health inequalities, psychotherapy and the evaluation of healthcare services. He has a particular interest in topics around mental health care.

ACKNOWLEDGEMENTS

Thanks to Richard D. Wiggins, Malcolm Williams and Betsy McCoach for inviting me to write this book. To Betsy McCoach for generously providing feedback on draft chapters. To the team at Sage for editorial support. To Miranda Ching and Sören Stöber for lending me a beautiful room to write in. To my colleagues for giving me time. To everyone I ever taught statistics for helping me learn. To Pippa Hembry for being there.

Thanks also to

- Gerbert Kraaykamp for kind permission to use data from the Family Survey Dutch Population;
- The Association of Religion Data Archives for making available data from the Faith Matters Survey, 2006;
- Alisha Holland for publishing the data from her study on The Distributive Politics of Enforcement;
- The Gapminder Foundation for making available data on life expectancy and GDP from around the world; and
- The UK Data Archive for permission to use data from the British Cohort Study 1970.

The data analyses reported in this book were conducted using the R Software for Statistical Computing (R Core Team, 2019) with the RStudio environment (RStudio Team, 2016). Some profile likelihood confidence intervals were calculated in Stata (StataCorp, 2019). All graphs were made in R, using the package ggplot2. Other R packages used in the making of this book are *brant, catspec, COUNT, Epi, gamlss, gapminder, grid, lmtest, MASS, mgcv, nnet, plyr, pscl, psych, reshape2, tidyverse,* and *VGAM.*

PREFACE

This is a book about regression models for categorical and count data, as they are used in the social sciences. It continues the previous book in the *SAGE Quantitative Research Kit* series, Volume 7, which introduced linear regression and the principles of statistical modelling. The book you are reading now shows how these principles can be applied when the outcome is a categorical or count variable. Reading Volume 7 of the *Kit* would be a good preparation for studying this book, but this book does stand on its own. Readers who are already familiar with linear regression can dive straight in. A brief reminder of linear regression is provided in Chapter 1, alongside an introduction to the mathematical notation used throughout.

Software

This book is software-neutral. It can be read and understood without using any statistical software. On the other hand, what you learn here can be applied using any statistical software that can estimate regression models. To carry out the analyses reported in this book, I mainly used the free open-source software R (R Core Team, 2019). On a few occasions, I also made use of the commercial software Stata (StataCorp, 2019). Other statistical packages often used by social scientists to estimate regression models are SPSS and SAS.

Web support pages with worked examples

It is generally a good idea to learn statistics by doing it – that is, to work with data sets and statistical software and play around with fitting statistical models to the data. To help with this, the support website for this book supplies data sets for most of the examples used in this book and gives worked examples of the analyses using all of the statistical models introduced in this book.

The support website is written in the R software. R has the advantage that it can be downloaded free of charge and that it has a growing community of users who write new add-on packages to extend its capability, publish tutorials and exchange tips and tricks online. However, if you prefer to use a different software, or if you are required to learn a different software for a course you are attending, you can download the data sets from the support website and read them into your software of choice. Instructions for this, as well as instructions on how to download R for free, are given on the support website. Head to

https://study.sagepub.com/quantitativekit

References

R Core Team. (2019). *R: A language and environment for statistical computing.* R Foundation for Statistical Computing. www.R-project.org/

RStudio Team. (2016). *RStudio: Integrated development for R.* RStudio. www.rstudio. com/

StataCorp. (2019). *Stata statistical software. Release 16.* Stata. www.stata-uk.com/

1

INTRODUCTION

Chapter Overview

This book is intended as a first introduction to regression models for outcomes that are either categorical or count variables. **Count variables** represent social phenomena that can be counted, such as the number of crimes in a city, the number of times a person visits a hospital, or the number of Members of Parliament who change party allegiance. **Categorical variables** represent social phenomena that cannot be measured numerically. We consider three types of categorical variables:

1 **Dichotomous variables** have exactly two categories. Examples are presence of an illness (the patient is either 'ill' or 'not ill') or retirement status ('retired' or 'not retired').
2 **Ordinal variables** have three or more categories that can be placed in a natural order. Examples are highest qualification ('no qualification', 'completed primary school', 'completed secondary school', 'university degree', etc.) or subjective health status based on an ordered response scale ('poor health', 'fair', 'good', 'very good health').
3 **Nominal variables** have three or more categories that cannot be placed in a meaningful order. Examples are choice of study subject ('science', 'humanities', 'arts', etc.) or type of accommodation ('rented', 'owned with mortgage', 'owned outright', 'nursing home or other institution', 'homeless').

The models discussed in this book include the following:

- **Logistic regression** for dichotomous (binary) outcomes
- The **general ordered logit model** for ordinal outcomes (also known as **ordinal logistic regression**)
- **Multinomial logistic regression** for nominal outcomes
- Several models for count outcomes, including **Poisson** and **negative binomial regression**, as well as **zero-truncated**, **zero-inflated** and **hurdle models**

This book assumes that the reader is familiar with linear regression, elementary **inferential statistics** (hypothesis tests and confidence intervals) and general methodological considerations in the collection and analysis of quantitative social science data. A good way to acquire or refresh this knowledge is to study the volumes in *The SAGE Quantitative Research Kit* series that precede this one. In particular, Volume 7 gives a thorough introduction to linear regression.

Why study regression models for categorical and count data?

Regression models are used widely in the social sciences to investigate relationships between social phenomena, to test theories about the social world, and to provide model-based predictions of what might happen in the future. Research examples discussed in Chapters 2 to 5 of this book include the following:

- *Health inequalities:* In England, people living in poorer areas are less likely to make use of free eye tests than people living in richer areas. Why?
- *Mental health:* Can we identify childhood experiences and characteristics that are associated with the risk of developing an eating disorder as an adult?
- *Sociology of culture:* Do people choose their cultural activities to display their social status?
- *Political science:* Under what circumstances are local politicians prepared to tolerate illegal street vendors in their cities?
- *Sociology of religion:* Is it true that people with a strong religious identity are more likely to be happy with their lives? And if so, why might that be?

This book does not provide conclusive answers to any of these questions. But it does discuss the methods used to investigate them.

A few words on terminology

Outcome and predictor

In regression models, we distinguish between the outcome variable and the predictor variables. The outcome variable is what we wish to explain or predict, the predictor variables contain the information that does the explaining or predicting. In different texts, you may find other names for these concepts:

- The **outcome variable** is also known as the dependent variable, or the response. It is usually denoted by the letter Y.
- **Predictor variables** are also known as independent variables, explanatory variables, or exposures. The conventional symbol for a predictor variable is X. When there are multiple predictor variables, they are identified by numeric subscripts: X_1, X_2, X_3, and so forth.

Types of variables

We can distinguish numeric and categorical variables. The values of a **numeric variable** are numbers that represent numeric measurements, such as a person's height or a country's gross domestic product. Among the numeric variables, we distinguish continuous and discrete variables:

- **Continuous variables:** A continuous variable is a numeric variable that can take any value within its possible range. For example, age is a continuous variable: a person can be 28 years old, 28.4 years old, or even 28.397853 years old. Age changes every day, every minute, every second, so our measurement of age is limited only by how precise we can or wish to be.

- **Discrete variables:** A discrete variable is a numeric variable that can only take particular, 'discrete' values. Count variables are discrete. They can take the values 0, 1, 2, 3 and so on. Consider the count variable 'number of children': you can have zero children, one child or seven children, but not 1.5 children.

In contrast to numeric variables, the values of categorical variables are not numbers, but categories. In a particular data set, the categories might be represented by numbers, but then the numbers are merely names for the categories and do not represent true numeric measurements. Three types of categorical variables were defined in the previous section.

Why do we need to look beyond linear regression?

When the intended outcome of an analysis is a categorical or count variable, linear regression is often not appropriate. With dichotomous or ordinal outcomes, a form of linear regression can sometimes be applied but, in general, is rarely advisable. (Some reasons are given in Chapter 2.) Nominal variables with three or more categories cannot meaningfully be modelled using linear regression at all. Linear regression models applied to count outcomes often fail to meet some of their assumptions. In particular, errors and residuals from a linear regression on count data are often not normally distributed and not **homoscedastic**.

So, to construct statistical models for categorical and count outcomes, we need specialised techniques. This book does not give a complete overview of all models that can be applied to categorical and count outcomes. But it does discuss the models most commonly used in the social sciences. Thus, for categorical outcomes, this book discusses models of the logistic regression family. Less frequently used alternatives (e.g. probit models, or the linear probability model for binary outcomes) are mentioned, but not discussed in detail. For count variables, Chapter 5 focuses on the most common forms of Poisson and negative binomial regression and discusses variants, including zero-truncation, zero-inflation and hurdle models. Other models for count data, such as quasi-Poisson, Poisson inverse Gaussian or less frequently used forms of negative binomial regression, are mentioned but not discussed in detail.

Regression beyond the linear model: an illustrated introduction

As a general introduction to the regression models discussed in this book, let's consider the ways in which they are the same as linear regression and the ways in which they are different. This section gives an intuitive introduction to these differences and the similarities. The subsequent section will be more precise and mathematical.

Consider Figure 1.1, which illustrates a linear relationship between two hypothetical continuous variables X and Y. This is an example for the sort of data where linear
regression may be appropriate. In particular,

- the outcome variable Y is continuous,
- the relationship between X and Y looks linear, and
- Y has about the same variation at all levels of X (homoscedasticity).

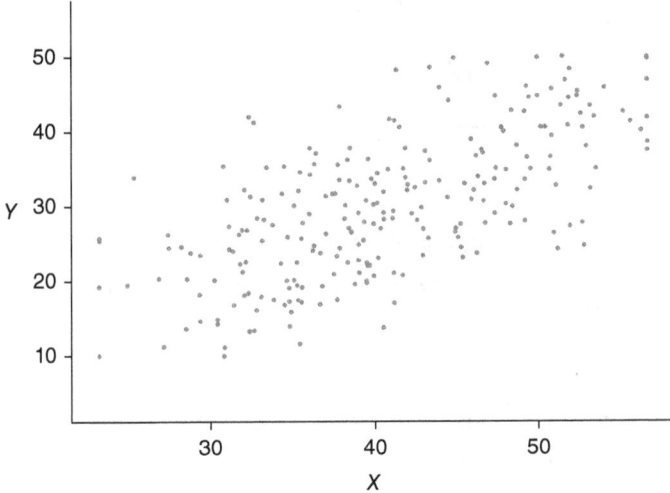

Figure 1.1 Hypothetical data illustrating a linear relationship between a continuous
outcome and a continuous predictor

Figure 1.2 illustrates the results of fitting a linear regression model of Y on X to these
data. The regression line shows the **predicted values** of Y, conditional on the values
of X. The errors around the prediction are assumed to be normally distributed with constant variance (homoscedasticity). This is illustrated by the three normal distributions
drawn around the predicted values of Y, for X values of 30, 40 and 50. Think of these
normal distributions as growing out of the page of this book towards you: the height
of the curve represents the probability of observing Y values in the range covered by
a section of the curve. These normal curves illustrate the predicted distribution of Y,
at each of the three selected values of X. Note that these three normal distributions
all have the same variance, representing the homoscedasticity assumption. (There is
nothing special about the X-values 30, 40 and 50; they are just chosen as an illustration. The errors from a linear regression are assumed to be normally distributed with
equal variance at *all* values of X. Thus, a three-dimensional representation of Figure 1.2
would resemble an elongate hill rising up from the page, its cross-section shaped like
the normal distribution, its top ridge sitting exactly over the regression line.)

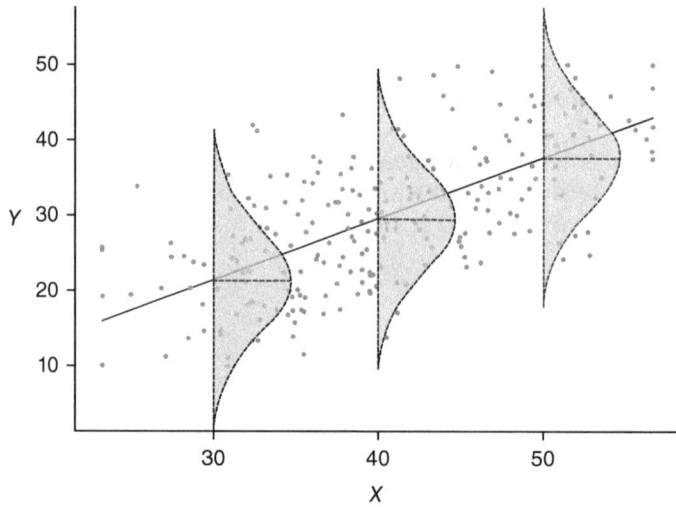

Figure 1.2   Illustration of a linear regression model

Note. This figure uses the hypothetical data from Figure 1.1. Think of the normal error distributions as growing straight out of the page towards you.

Now consider Figure 1.3. This illustrates data for which some important assumptions of linear regression are not met. In particular,

- the outcome variable Y is not continuous but discrete, taking only the values 0, 1, 2 and so forth;
- the relationship between X and Y does not look linear; and
- the variance of Y differs depending on the values of X: at small values of X, there is less variation in Y than for large values of X (**heteroscedasticity**).

A model for the data in Figure 1.3 is illustrated in Figure 1.4. This is an example of a Poisson regression (see Chapter 5). The black curve represents the predicted values of Y, given the values of X. The grey bars represent the predicted distribution of Y, for X-values of 30, 40 and 50. Think of the grey bars as growing out of the page towards you. The height of the bar represents the probability of observing each value of Y, given an X-value. Note the following properties of this model:

- *Discrete predicted distribution*: The predicted distribution of Y, contingent on the values of X, takes only the discrete values 0, 1, 2, ..., like Y itself.
- *Non-linear relationship*: The curved regression line represents the curvilinear relationship between X and Y assumed by this model.
- *Heteroscedasticity*: The predictive distributions at $X = 30$, 40 and 50 have different variances. When the predicted value of Y is small, so is the variance. For larger predicted values of Y, the variance is also large. So, unlike in a linear regression, in this regression model the variance is assumed to depend on the predicted value.

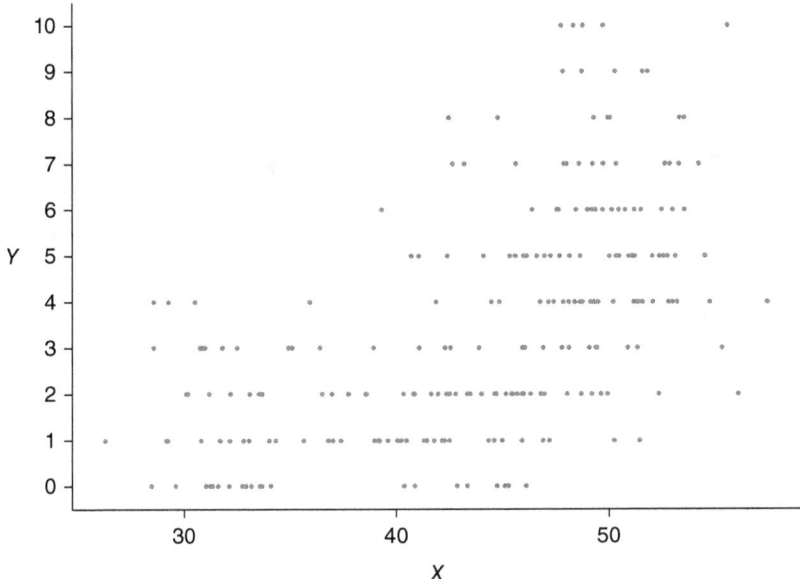

Figure 1.3 Hypothetical data illustrating the relationship between a discrete outcome and a continuous predictor

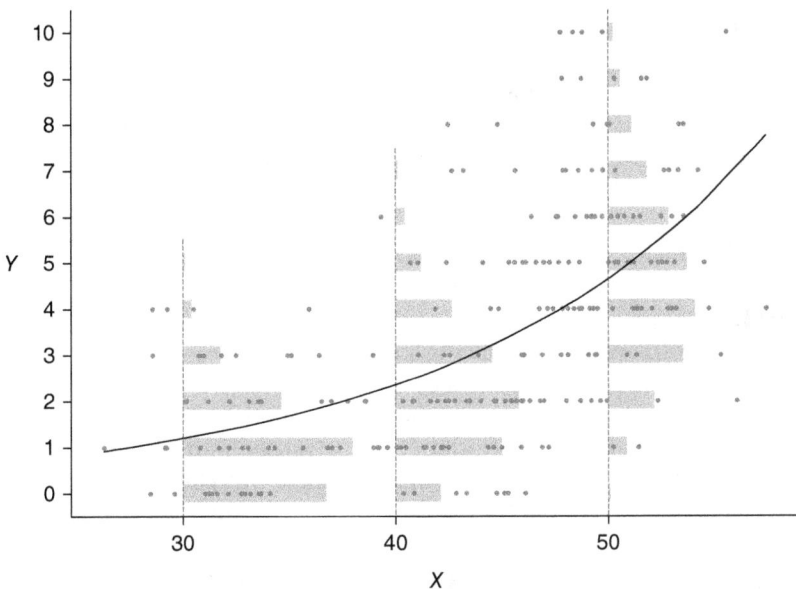

Figure 1.4 Illustration of a Poisson regression model

Note. This figure uses the hypothetical data from Figure 1.3. Think of the grey bars as growing out of the page towards you. These grey bars represent the probability of observing $Y = 0$, $Y = 1$, $Y = 2$ and so forth, conditional on the values of X.

Linear regression: a reminder, with some mathematical notation

Let's briefly revisit linear regression in a bit more detail. Readers who are confident in their knowledge of linear regression may wish to skip this section. However, this section also introduces some elements of mathematical notation that may be slightly different (although not very different) to the notation used in other books or sources. Therefore, even if you are a seasoned user of linear regression, you may at least want to have a quick glance at this section. The notation used in this book is consistent with *The SAGE Quantitative Research Kit*, Volume 7.

Regression model and notation

Linear regression is a model of the form

$$Y_i = \beta_0 + \beta_1 X_{1i} + \beta_2 X_{2i} + \ldots + \beta_k X_{ki} + \varepsilon_i$$

$$\varepsilon_i \sim N\left(0, \sigma^2\right)$$

where

- Y is a continuous outcome variable, and Y_i is the value of the outcome variable for the ith person.
- $i = 1, 2, 3, \ldots, n$ is an identifier for members of our data set, numbered from 1 to n, where n is the sample size. For example, Y_1 denotes the outcome value of the first member of our data set, Y_{143} denotes the outcome value for the 143rd member, and so forth.
- $X_1, X_2, \ldots, X_k$ are predictor variables, numbered from 1 to k, where k is the number of predictor variables in the model. X_{1i} denotes the value of X_1 for the ith person.
- β_0 is the **intercept**; this is the predicted value of Y when all predictors are zero.
- $\beta_1, \beta_2, \ldots, \beta_k$ are **slope** coefficients that relate the predictors to the outcome.
- ε_i are called the errors; the error ε_i is the difference between the actual outcome value Y_i and the outcome value predicted by the model.
- $\varepsilon_i \sim N(0, \sigma^2)$ is statistical notation indicating the assumption that 'the errors are normally distributed, with mean zero and variance σ^2.' The tilde symbol (~)means 'is distributed as', and N denotes the normal distribution.

The regression equation can be used to obtain a predicted value of Y. We denote a predicted value by $\hat{Y}$ (pronounced 'Y-hat'). The hat (^) signifies a prediction from a statistical model, in contrast to an observed value in a data set. Predictions from a linear regression are calculated by the predictive equation

$$\hat{Y}_i = \beta_0 + \beta_1 X_{1i} + \beta_2 X_{2i} + \ldots + \beta_k X_{ki}$$

It follows that the outcome is represented in a statistical model as the sum of a predicted value and an error:

$$Y_i = \hat{Y}_i + \varepsilon_i$$

Errors and residuals

The errors are the differences between the observed values and the values predicted by the statistical model. Rearranging the last equation from the previous section, we can express this mathematically as

$$\varepsilon_i = Y_i - \hat{Y}_i$$

Some books and sources may use the term *residual* to refer to the errors. In this book, as in Volume 7 of *The SAGE Quantitative Research Kit*, I make a distinction between the errors and the residuals.

- **Errors** are the differences between observed values of Y and the predicted values $\hat{Y}$, when these predicted values come from the 'true' regression line. The true regression line is that which would result if I had complete information (e.g. population data) on the relationship between Y and its predictors, such that I could know the true values of the **coefficients** $\beta_0, \beta_1, \ldots$ in the regression equation.
- **Residuals** are what I observe in practice when I fit a linear regression model to a data set. They are the differences between the observed values Y and the predicted values $\hat{Y}$, when these predicted values come from a regression line estimated on a sample of data. Here, the prediction of $\hat{Y}$ is based not on the true coefficients $\beta_0, \beta_1, \ldots$, but instead on their **estimates** from a particular data set. We denote the estimated coefficients using the 'hat notation' by $\hat{\beta}_0, \hat{\beta}_1, \ldots$ We can think of the residuals as estimates of the errors, which we obtain by fitting a model to a particular data set.

Estimation and partition of outcome variance

Using data to estimate the coefficients of a regression model is called **fitting the model** or *estimating the model on a data set*. The coefficients of a linear regression model can be estimated from the data via a mathematical procedure called least squares estimation. This finds the coefficients that minimise the sum of the squared residuals. Due to the estimation procedure involved, linear regression is also called *ordinary least squares* (OLS) regression.

OLS regression involves partitioning the total variation of the outcome into two parts: (1) the variation that is related to (can be 'explained by') the predictor variables and (2) the residual variation that is unrelated to the predictor variables. The proportion of the total outcome variation explained by the predictors of a linear regression model is measured by the R^2 **(R-squared)** statistic, also called the *coefficient of determination*. The R^2 statistic is often used as an indicator of model fit, or **goodness of fit**. Different regression models, if they are nested, can be compared in terms of how much of the outcome variance they respectively explain, and the F-test for model comparison can be used to formally test the null hypothesis that the larger of two nested models does not explain a greater proportion of the outcome variance than the smaller model.

Generalised linear models

Now let's consider how we might move from linear regression to a more general way of representing regression models, including regression models for categorical and count outcomes. The **generalised linear model** (GLM) is a type of statistical model that encompasses, as special cases, linear regression as well as many (but not all) of the models discussed in this book. In particular, logistic regression, ordinal and multinomial logistic regression, Poisson regression and some variants of negative binomial regression are special cases of the GLM.

To understand how GLMs work, it is helpful to write the linear regression model in a slightly different way:

$$\mu_i = \beta_0 + \beta_1 X_{1i} + \beta_2 X_{2i} + \ldots + \beta_k X_{ki}$$

$$Y_i \sim N\left(\mu_i, \sigma^2\right)$$

Here, μ_i is the mean of the outcome for sample members with certain characteristics represented by the predictors $(X_{1i}, X_{2i}, \ldots, X_{ki})$. Our outcome Y_i follows a normal distribution with mean μ_i, where the mean depends on the predictors $X_1, X_2, \ldots, X_k$. We say that the outcome follows a normal distribution with the mean *conditional* on the predictors. Note that the variance σ^2 does not have an index i. That is, the variance is assumed to be the same for all Y_i and does not depend on the predictors. This is the assumption of homoscedasticity.

When we view the linear regression model as an instance of the GLM, then two things become a matter of choice:

1   The outcome on the left-hand side of the equation may be transformed in some way via a mathematical function, such as the logarithmic function or the inverse

function. In the context of a GLM, a function to transform the outcome is called a *link function* (because the outcome is *linked* to the predictors via this function).

2   The distribution of the outcome variable (with mean conditional on the values of the predictors) may be the normal distribution or a different distribution from a general type called the exponential family of distributions.

For example, the Poisson regression model (as illustrated in Figure 1.4) can be written as

$$\log(\mu_i) = \beta_0 + \beta_1 X_{1i} + \beta_2 X_{2i} + \ldots + \beta_k X_{ki}$$

$$Y_i \sim Poisson(\mu_i),$$

Here, a logarithmic link function is employed, so the predictors are related to the logarithm of the mean rather than directly to the mean itself. The outcome is assumed to follow a Poisson distribution (rather than a normal distribution), with mean conditional on the values of the predictors. It is a property of the Poisson distribution that the mean is equal to the variance. This implies that the variance, like the mean, depends on the values of the predictors, and so the observed values around the Poisson regression line are assumed to be heteroscedastic (not homoscedastic), as we saw above in the section 'Regression Beyond the Linear Model'.

When fitting a GLM to a data set, in general the coefficients are found not via OLS (as in linear regression) but via a procedure called maximum likelihood estimation.[i] Details of maximum likelihood estimation are beyond the scope of this book, but an intuitive explanation will be provided in Chapter 2. With maximum likelihood estimation, there is no partitioning of the outcome variance as in OLS regression, and consequently neither the R^2 statistic nor the F-tests of model comparisons are suitable for evaluating models estimated by maximum likelihood. As we will see, in the models discussed in this book, we will use statistics other than R^2 to measure model quality and use statistical tests called likelihood ratio tests for model comparisons, instead of the F-test.

What's the same and what's different

Although the models discussed in this book differ from linear regression in many respects, the principles of statistical modelling that apply in linear regression (and

---

[i]Linear regression can be performed either via OLS or via maximum likelihood. The two procedures result in the same coefficient estimates. However, for other GLMs OLS estimation is not suitable.

which were introduced in Volume 7 of *The SAGE Quantitative Research Kit*) also apply to all the models discussed here. What differs is the specific techniques used to put these modelling principles into practice. For example, every statistical model relies on assumptions, but different models make different assumptions, and the techniques for investigating the plausibility of assumptions differ, too. In applying any type of model, we are concerned with goodness of fit and predictive power, but the measures of these concepts are not the same across all model types. When we are interested in formally comparing two alternative models, the appropriate statistical test or index for doing so depends on the types of models we wish to compare. And in any statistical model discussed in this book, we are interested in interpreting slope coefficients of predictors, but our ways of interpreting them will again differ depending on the type of model we employ.

As we have seen, some assumptions made in linear regression are not made by the models discussed in this book. In particular, the assumptions of normality, homoscedasticity and linearity do not apply. However, other assumptions made in linear regression models also hold for the models discussed in this book. In particular:

Randomness of errors: *The errors are the result of a random process.* If the errors are not in fact random, then **inferences** from the sample to a population or process might be illusory. A random process may be built into the design of our study, such as when we employ random sampling to select survey respondents. But we may also assume randomness in social or natural processes.

Independence of errors: *The errors are independent of one another.* If there is dependency between some or all errors, hypothesis tests and confidence intervals around model coefficients and model predictions may yield misleading results.

The predictor is measured without error. GLMs have no random terms for the predictor variables, and thus implicitly assume that the predictor variables are measured without error. Structural equation models (see *The SAGE Quantitative Research Kit, Volume 9*) are one way to analyse data taking into account measurement errors in both predictor and outcome variables.

Absence of extremely influential observations: There are no extreme outliers in the data that distort the observed relationship between the predictor and the outcome. If this assumption is not met, coefficient estimates, as well as hypothesis tests and confidence intervals, may be misleading.

Absence of collinearity: Collinearity occurs when one or more predictor variables are perfectly or almost perfectly predictable from one or several other predictors. When there is perfect collinearity, the estimation of the regression model will break down. A computer program faced with perfect collinearity among model predictors will either return an error or automatically omit one of the variables that cause collinearity.

When predictors are very highly correlated, even though not perfectly collinear, the model estimation may be mathematically possible, but the results might be unstable. 'Unstable' here means that the coefficient estimates would be liable to change drastically in response to small changes in the data. This can sometimes be recognised if, after estimation, some coefficients turn out to have very large standard errors.

How you might use this book

If you are new to regression models for categorical and count data, it is best to study Chapter 2 before Chapters 3 to 5. There are several reasons for this. Chapter 2 covers logistic regression for binary outcomes and, in doing so, introduces essential concepts such as probability, odds and the logit transformation, which are fundamental to ordinal and multinomial logistic regression (Chapters 3 and 4) also. Some general techniques, such as likelihood ratio tests and profile likelihood confidence intervals, which are relevant in all subsequent chapters, are also first discussed in Chapter 2.

On the other hand, Chapters 3 to 5 can be read in any order. You should not have a problem, for example, if, having studied binary logistic regression in Chapter 2, you jump straight to models for count regression in Chapter 5. Similarly, you may choose to read Chapter 4 (on multinomial logistic regression) before Chapter 3 (ordinal logistic regression).

The last chapter in this book, Chapter 6, does not focus on a particular model type, but rather on the practice of statistical modelling itself. It addresses the question how we can give ourselves the best chance of arriving at a 'good model' for our data, and how we can avoid common pitfalls, such as overfitting, underfitting and misuse of statistical hypothesis tests, which are responsible for many erroneous scientific 'findings' published in reputable journals (and sometimes lavishly reported in the media). Chapter 6 assumes that the reader is familiar with essential modelling techniques, so it is probably recommendable to read at least Chapter 2 before embarking on Chapter 6, unless you already have some knowledge and practical experience of statistical modelling.

Further Reading

Krzanowski, W. J. (1998). *An introduction to statistical modelling*. Wiley.
There is a general mathematical formulation of the generalised linear model, but it is beyond the scope of this book to explain it, and it's not necessary to understand for the purposes of this book. If you are interested, this text is a good place to start.

2

LOGISTIC REGRESSION

Chapter Overview

What is logistic regression?

Logistic regression is a statistical model for a *dichotomous* outcome. 'Dichotomous' means that the outcome has exactly two categories. Some examples are:

- Presence of an illness (the patient is either 'ill' or 'not ill')
- Outcome of surgery ('survival' or 'death')
- Retirement status ('retired' or 'not retired')

A synonym for dichotomous in this context is **binary**.

Logistic regression models the *probability* of an *event*, such as the probability that a person will become ill, the probability that a person will survive surgery, or the probability that a person will retire by the age of 60.

It may help to emphasise, at this point, some fundamental differences between linear regression and logistic regression. Linear regression models a *numeric* outcome and investigates how the *mean* of this outcome varies depending on the values of a set of predictors. An estimate of a mean in a population can be derived via the *mean of the outcome in a random sample* of people taken from this population. In contrast, logistic regression models a dichotomous outcome and investigates how the *probability* of the outcome event varies depending on the values of a set of predictors. An estimate of the probability of an event in a population can be derived from the *proportion of the members of a random sample for whom the outcome event occurs*. There are other differences between linear and logistic regression, and we shall discuss many of them in this chapter.

Logistic regression may be used for the following purposes:

- *Understanding relationships:* Researchers may be interested in how a set of variables relates to a dichotomous outcome to address substantive research questions. For example, logistic regression has been employed to investigate how the risk of developing an eating disorder as an adolescent or a young adult is associated with childhood characteristics, such as family background, weight and eating habits (Nicholls et al., 2016).
- *Investigating theories:* Logistic regression might be employed to investigate whether the relationships between a dichotomous outcome and a set of predictor variables are consistent with social science theories. For example, Kraaykamp et al. (2010) were interested in theories of cultural consumption and posited that people in high social class positions are motivated to attend 'highbrow' cultural events, such as classical concerts, as a way of displaying their social status. Kraaykamp et al. investigated their theory by modelling the relationship between social status and the probability of attending highbrow cultural events, while controlling for indicators of other possible reasons why high-status individuals might attend, or can afford to attend, such events, such as income and education. (I will use Kraaykamp et al.'s data later in this chapter, and also again in Chapter 4.)

- *Risk prediction:* Logistic regression might be used to build a predictive model that can estimate the probability of an event occurring in the future. For example, surgeons may wish to predict the probability that a patient, who is considering surgery, will not survive the procedure. This probability may be predicted from characteristics of the patient, such as their age, the severity of their disease, the presence of comorbid diseases, and so forth. Risk prediction may aid in the decision whether surgery is the best treatment or may help to plan post-surgery care in the hospital (Eugene et al., 2018).

Logistic regression is also important because it is used as an element within more complex methods, such as zero-inflated and hurdle models (Chapter 5). Logistic regression is sometimes called 'binary logistic regression' to distinguish it from related methods, such as ordinal logistic regression and multinomial logistic regression, which are discussed in Chapters 3 and 4 of this book, respectively.

To understand logistic regression, it is necessary to equip ourselves with several concepts that are commonly used in the description and analysis of dichotomous outcome variables. The following three sections will discuss these essential concepts – namely, probability, conditional probability, odds and odds ratios. If you are already familiar with all of these, you may wish to jump straight to the 'Log Odds' section, which defines the logit transformation that is used in logistic regression.

Probabilities and conditional probabilities

In logistic regression, we are interested in modelling the probability of an event. A probability can take values between 0 and 1, where 0 means that the event never occurs, while 1 means that the event is certain to occur. A probability of 0.5 indicates that we expect the event to occur half of the time. If the probability of getting a cold this month is 0.03, this means that we expect about three people in a hundred to get a cold.

It will be useful to introduce some notation. Let's call the outcome Y. This is the same symbol commonly used for the outcome in a linear regression (e.g. as in *The SAGE Quantitative Research Kit*, Volume 7). However, in linear regression, Y is a continuous variable and can, in theory, take many different values. In logistic regression, Y is defined as a **binary variable** and can take only two values: either $Y = 1$ or $Y = 0$.

- When $Y = 1$, we call this an *event*. For example, the event may be 'the person is ill'.
- When $Y = 0$, we call this a *non-event*, such as 'the person is not ill'.

We will write the probability of the event occurring like this:

$P(Y = 1)$

The probability of a non-event is written as

$$P(Y = 0)$$

Because Y can only take the values 0 or 1, the probabilities for these two values must sum to 1:

$$P(Y = 1) + P(Y = 0) = 1$$

It follows that, once we know the probability of the event, we also know the probability of the non-event, because

$$P(Y = 0) = 1 - P(Y = 1).$$

To simplify the notation, we also sometimes denote the probability of the event by the letter p. That is,

$$p \equiv P(Y = 1)$$

where the symbol $\equiv$ means 'is defined as'.

In statistical modelling, we are usually interested in how probability of an event depends on other variables. This is called the *conditional probability* of an event. For example, a person's gender may be related to the risk of getting a specific disease. Breast cancer is rarer in men than in women. So the probability of the event 'breast cancer' is conditional on the variable 'gender'. We write a conditional probability as follows:

$$P(Y = 1 | X = x)$$

This denotes the conditional probability of the event '$Y = 1$' given the value x of a predictor variable X. For example, the event may be 'the person is ill', and the predictor X may be gender, coded as male or female. Then we can write the probability of illness for females as follows:

$$P(Y = 1 | \text{Gender} = \text{Female})$$

Simple example of data with a binary outcome

Let's examine a simple example of data that can illustrate these concepts. Table 2.1 shows data from the Family Survey Dutch Population (FSDP), conducted in 1998 (de Graaf et al., n.d.; Kraaykamp et al., 2010). This shows the relationship between attendance at classical concerts and level of education. Classical concert attendance was assessed by a survey question that asked respondents whether they attended at least one classical concert per year.

Table 2.1 Numbers (and percentages) of attendance at classical concerts, by education

		Education					
		Below High School		High School or Higher		Total	
Attends classical concerts	Yes	87	18.5%	137	55.5%	224	31.2%
	No	383	81.5%	110	44.5%	493	68.8%
	Total	470	100.0%	247	100.0%	717	100.0%

Note. Data were taken from from the Family Survey Dutch Population (de Graaf et al., n.d.).

We may want to use the data in Table 2.1 to estimate the proportion of Dutch adults who attend classical concerts, separately for each of two educational groups. So we are interested in the *conditional probability* of a person attending classical concerts, given their education.[1] If we assume that the data in Table 2.1 come from a random sample of the Dutch population, then a plausible estimate of this probability is the proportion of classical concert attenders among the high school educated in our sample:

$$\hat{P}(\text{Classical} \mid \text{Education} = \text{'High school'}) = \frac{137}{137 + 110} = 0.555$$

We write $\hat{P}$ – that is, we put a 'hat' on the letter P – to indicate that this is an estimated probability, derived from sample data.

Similarly, we can also calculate an estimate of the conditional probability of classical concert attendance given no high school education (that is, for people whose highest qualification is lower than high school):

$$\hat{P}(\text{Classical} \mid \text{Education} = \text{'No high school'}) = \frac{87}{87 + 383} = 0.185$$

These data suggest that those with a high school education have a higher probability of attending classical concerts than those without a high school education. We will now use these data further to illustrate three other concepts that are important in logistic regression: the **odds**, the **log odds** and the **odds ratio**.

---

[1]Just to avoid misunderstandings: *probability of attending classical concerts* here strictly means the probability of responding 'I attend classical concerts at least once a year' to the survey question. From the data in Table 2.1, we cannot, for example, calculate the probability that a person will buy a ticket to a particular classical concert, or that they will attend a classical concert next year. Such events can be modelled too, but to do so, we would need different sorts of data.

Analysis of a 2 × 2 table: probabilities, odds and odds ratios

In the previous section, we considered a 2 × 2 table of classical concert attendance by educational level. Let's make our analysis more general. Consider a 2 × 2 table of the form given in Table 2.2.

Table 2.2 A generic 2 × 2 table

		Predictor		
		Group 1	**Group 2**	**Total**
Outcome	Event	a	b	$a + b$
	Non-event	c	d	$c + d$
	Total	$n_1 = a + c$	$n_2 = b + d$	$N = n_1 + n_2$

Let's assume that the data in this table (represented by the quantities a, b, c and d, and the resulting totals) come from a random sample drawn from some population, and that we wish to use these quantities to estimate how the probability of the event varies by group membership.

Probabilities

The predicted probability of an event for members of Group 1 is estimated as follows:

$$\hat{P}(\text{Event} \mid \text{Group1}) = \frac{a}{a+c} = \frac{a}{n_1}$$

The predicted probability of an event for members of Group 2 is estimated as follows:

$$\hat{P}(\text{Event} \mid \text{Group2}) = \frac{b}{b+d} = \frac{b}{n_2}$$

These are the same calculations we used above to derive the predicted probability of attending classical concerts from Table 2.1.

Risk Ratio and Absolute Risk Difference

In research, we are often interested in comparing two groups with respect to the probability of an event. Two common measures of such a difference between groups are the risk ratio and the absolute risk difference.

The risk ratio (RR) is calculated as follows:

$$RR = \frac{P(\text{Event} \mid \text{Group1})}{P(\text{Event} \mid \text{Group2})} = \frac{a/n_1}{b/n_2}$$

This gives us a measure of how much greater the probability (or 'risk') of the event is in Group 1 compared to Group 2. For example, using the data in Table 2.1, we might estimate the risk ratio for classical concert attendance by education (high school vs no high school) as follows:

$$\widehat{RR} = \frac{\hat{P}(\text{Classical} \mid \text{High school})}{\hat{P}(\text{Classical} \mid \text{No high school})} = \frac{137/247}{87/470} = \frac{0.555}{0.185} = 3.00$$

So the probability of attending classical concerts is estimated to be three times as high among the high school educated compared to those with lower than high school education.

The absolute risk difference (ARD) is calculated simply by subtracting one probability from another:

$$ARD = P(\text{Event} \mid \text{Group1}) - P(\text{Event} \mid \text{Group2})$$

Using again the data from Table 2.1, we can estimate the absolute risk difference as follows:

$$\widehat{ARD} = \hat{P}(\text{Classical} \mid \text{High school}) - \hat{P}(\text{Classical} \mid \text{No high school})$$
$$= 0.555 - 0.185 = 0.370$$

So the absolute difference in the probability of classical concert attendance between those with high school education compared to those without is 0.37, or 37 percentage points.

The risk ratio and the absolute risk difference are both valid measures of a between-group difference in the probability of an event. However, as we will see, logistic regression makes use of a different measure of this difference, namely the odds ratio. We will discuss the odds ratio presently. Before we do so, we first need to understand the concept of the odds.

Odds

The odds denote the number of times an event occurs divided by the number of times an event does not occur. If you have ever placed a bet with a professional bookmaker, you have encountered odds. For example, if the odds of a horse winning the next race are 7/5, this means that the horse is expected to win this sort of race 7 out of 12 times, and lose the race 5 out of 12 times. In betting, odds are used to calculate the winnings for a successful bet. But the odds are also an important concept in research.

Referring again to Table 2.2, the odds of the event for Group 1 are estimated as follows:

$$\widehat{Odds}(\text{Event} \mid \text{Group 1}) = \frac{a}{c}$$

And for Group 2:

$$\widehat{Odds}(\text{Event} \mid \text{Group 2}) = \frac{b}{d}$$

So when computing the estimated odds, we are dividing the number of people with the event by the number of people without the event. This is in contrast to calculating probabilities, where we divided the number of people with the event by the total number of people.

Using the data in Table 2.1, the estimated odds of classical concert attendance for those with high school education and those without high school education are, respectively:

$$\widehat{Odds}(\text{Classical} \mid \text{High school}) = \frac{137}{110} = 1.25$$

$$\widehat{Odds}(\text{Classical} \mid \text{No high school}) = \frac{87}{383} = 0.23$$

Odds can take values between 0 to $+\infty$. This is in contrast to probabilities, which can take values between 0 and 1.

We can also define the odds relative to probabilities. Consider the following calculation:

$$\widehat{Odds}(\text{Event} \mid \text{Group 1}) = \frac{a}{c} = \frac{a/n_1}{c/n_1} = \frac{a/n_1}{1 - a/n_1}$$

$$= \frac{\hat{P}(\text{Event} \mid \text{Group 1})}{1 - \hat{P}(\text{Event} \mid \text{Group 1})}$$

It follows that the odds can also be written (with simplified notation) as follows:

$$Odds = \frac{p}{1-p}$$

where p is the probability of the event. This will be important soon, when we see how the logistic regression model is defined.

Log odds: The logit transformation

The log odds are defined as the natural **logarithm** of the odds:

$$log\ odds = log\left(\frac{p}{1-p}\right)$$

If you need to remind yourself what logarithms are, or what a natural logarithm is specifically, please consult Box 2.1.

Box 2.1

Logarithms

A logarithm is a mathematical function that has applications in many areas of science, including statistics. Logarithms are used to transform the outcome variable in all the models discussed in this book. So to understand this book, it is essential to understand logarithms.

Most people, when they first encounter logarithms, learn about base-10 logarithms. A base-10 logarithm of a number n is the answer to the question: '10 to the power of what equals n?'

We write $log_{10}(n)$ to denote the logarithm of n to base 10. For example,

- $log_{10}(100) = 2$, because $10^2 = 100$
- $log_{10}(1000) = 3$, because $10^3 = 1000$
- $log_{10}(0.1) = -1$, because $10^{-1} = 0.1$

One scientific application of logarithms is the Richter scale for measuring the strength of earthquakes, which uses base-10 logarithms. This means that the Richter scale is a *multiplicative scale*: an earthquake of strength 3 on the Richter scale is 10 times as strong as an earthquake of strength 2, which in turn is 10 times as strong as an earthquake of strength 1.

In general, we write $log_b(n)$ to denote the logarithm of n to base b. Any positive number can be a logarithmic base. Thus, $log_2(n)$ is the base-2 logarithm. For example,

- $log_2(8) = 3$, because $2^3 = 8$, and
- $log_2(0.5) = -1$, because $2^{-1} = 0.5$.

(Continued)

The logarithm is only defined for positive numbers. That is, $\log_b (0)$ has no result, and neither does $\log_b (-1)$. The following results hold for any base b:

- $\log_b (b) = 1$
- $\log_b \left(\dfrac{1}{b}\right) = -1$
- $\log_b (1) = 0$

An important type of logarithm is the *natural logarithm*, whose base is Euler's number $e = 2.71828....$ The natural logarithm has useful mathematical properties, and is therefore commonly used in many statistical techniques, including in all the models discussed in this book. Since the natural logarithm is the most important for mathematicians, it is assumed that if someone writes $\log(n)$ without specifying the base, the natural logarithm is meant. In logistic regression, the log odds of the outcome event are calculated using the natural logarithm. Table 2.3 gives some example values for odds and corresponding log odds.

Table 2.3 Comparing the scales of probabilities, odds and log odds

Probability	Odds	Log Odds
0	0	$-\infty$
0.01	$\frac{1}{99}$	−4.595
0.1	$\frac{1}{9}$	−2.197
0.25	$\frac{1}{3}$	−1.099
0.33333	$\frac{1}{2}$	−0.693
0.5	1	0
0.66667	2	0.693
0.75	3	1.099
0.9	9	2.197
0.99	99	4.595
1	$+\infty$	$+\infty$

Note. Log odds are given to three decimal places.

The transformation of probabilities into log odds is called the **logit transformation**. The log odds have a theoretical range from $-\infty$ to $+\infty$. As we will see presently, their unrestricted range makes the log odds useful in modelling a binary outcome variable. To see how probabilities, odds and log odds relate to one another, consider Table 2.3. In particular, note the following:

- When the probability $p = 0.5$, the odds are equal to 1, and the log odds are equal to 0.
- When $p < 0.5$, the odds are smaller than 1, and the log odds are negative.
- When $p > 0.5$, the odds are larger than 1, and the log odds are positive.

Odds Ratio

We saw earlier that a between-group difference in the probability of an event can be described by the risk ratio or by the absolute risk difference. Another relevant measure of the difference between two groups is the odds ratio. As we will see, the odds ratio is the usual measure of the **effect size** in the context of logistic regression, so it is of primary importance in this chapter.

We denote the odds ratio by OR, and an estimated odds ratio from observed data by $\widehat{OR}$. We calculate the estimated odds ratio as follows (referring once again to the generic Table 2.2):

$$\widehat{OR} = \frac{\widehat{Odds}(\text{Event} \mid \text{Group 1})}{\widehat{Odds}(\text{Event} \mid \text{Group 2})} = \frac{a/c}{b/d} = \frac{ad}{bc}$$

Using the numbers in Table 2.1 as an example, the observed odds ratio of classical concert attendance for high school education versus no high school education is as follows:

$$\widehat{OR} = \frac{\widehat{Odds}(\text{Classical} \mid \text{High school})}{\widehat{Odds}(\text{Classical} \mid \text{No high school})} = \frac{137 \times 383}{110 \times 87} = 5.48$$

The odds ratio quantifies the difference in classical concert attendance between these two groups of people. This result may be formally reported as follows: 'In this data set, people who completed high school have 5.48 times higher odds of attending classical concerts than people who did not complete high school'.

The odds ratio is a measure of the relationship between a predictor and a binary outcome. Theoretically, the odds ratio can assume values between 0 and $+\infty$. Negative odds ratios do not exist. The following may help in interpreting what a given odds ratio means:

- $OR = 1$ means that there is no relationship between the predictor and the binary outcome. For example, if the odds ratio for classical concerts comparing high school completers with others was 1, this would mean that there was no difference in the odds of classical concert attendance between the two groups.

- $OR = 2$ means that the odds of the event in one group are twice as high as the odds of the other group.
- $OR = 0.5$ means that the odds of an event for one group are half those of the other group.

Note that an OR of size S signals the same effect size as an OR of size $\frac{1}{S}$, but in reverse. Thus, the following two statements mean the same thing:

- The odds ratio for illness for males versus females is 2 (males have twice the odds of being ill compared to females).
- The odds ratio for illness for females versus males is 0.5 (females have half the odds of being ill compared to males).

An OR of 1.5 is sometimes reported as 'Group 1 has 50% higher odds of the event than Group 2'. Similarly, an OR of 0.67 is sometimes reported as: 'Group 2 has 33% lower odds of the event than Group 1' (since $1 - 0.67 = 0.33 = 33\%$).

Taking again Table 2.1 as an example, instead of calculating the odds ratio of classical concert attendance for the highly educated versus others, as we did above, we might calculate the odds ratio the other way around:

$$\widehat{OR} = \frac{\widehat{Odds}(\text{Classical} \mid \text{No high school})}{\widehat{Odds}(\text{Classical} \mid \text{High school})} = \frac{87 \times 110}{383 \times 137} = 0.18$$

We may report this as follows: the odds of classical concert attendance among those with no high school education are smaller than the odds of those with high school education by a factor of 0.18. Alternatively, you may find this result reported as follows: the odds of classical concert attendance are 82% lower for those without high school than for those with high school (since $1 - 0.18 = 0.82 = 82\%$).

Logistic regression: the model

We are now ready to define the logistic regression model. I give the general definition first, and then introduce a series of examples that, step by step, will illuminate logistic regression models and their interpretation.

Logistic regression models the logarithm of the odds of the outcome. The general logistic regression model is

$$\log\left(\frac{p_i}{1 - p_i}\right) = \beta_0 + \beta_1 X_{1i} + \beta_2 X_{2i} + \cdots + \beta_k X_{ki}$$

where we can understand the left-hand side of the equation, $\log\left(\frac{p_i}{1 - p_i}\right)$, as follows:

- p_i is the *probability* of the outcome event for the *i*th individual – that is, $p_i = P(Y_i = 1)$.
- $\dfrac{p_i}{1-p_i}$ are the *odds* of the outcome event for the *i*th individual.
- $\log\left(\dfrac{p_i}{1-p_i}\right)$ are the *log odds* of the outcome event for the *i*th individual. This is called the *logit transformation*. In the terminology of the generalised linear model, the logit transformation is the *link function* (see Chapter 1) employed in logistic regression.

The right-hand side of the logistic regression equation is of the same form as in linear regression (see Chapter 1), such that

- $X_{1i}, X_{2i}, \ldots, X_{ki}$ are predictors, which may be numeric or dummy variables, with values of the predictors varying by individual (*i*);
- $\beta_1, \beta_2, \ldots, \beta_k$ are slope coefficients corresponding to the predictors, and
- β_0 is the intercept.

So a difference between linear and logistic regression is that in logistic regression we predict the log odds of the outcome event, whereas in linear regression we predict the mean of the numerical outcome. This has implications for the interpretation of the coefficients: In logistic regression, the slope coefficients $\beta_1, \beta_2, \ldots, \beta_k$ are to be interpreted as the effects of the predictors on the log odds of the outcome event. The intercept, β_0, gives the log odds of the outcome event when all predictor variables are equal to zero.

Why do we use the logit transformation? Instead of the complicated-looking log odds, could we not just choose to set up the model so that we predict the probability that $Y = 1$? In fact, such a model does exist. It is called the linear probability model, and it can be useful in some situations. However, the linear probability model has an important disadvantage. Consider that a probability is a number that can take values between 0 and 1. In practice, if we chose to define the left-hand side of our model as the probability of the event, the model estimation may result in predicted values larger than 1 or smaller than 0, which are outside the range on which probabilities are defined. Thus, our predictions could potentially be nonsensical. Logistic regression avoids this disadvantage, because its outcome, the log odds, can take values between $-\infty$ and $+\infty$. This means that logistic regression will never result in predictions that are outside the theoretical range of its outcome.

A disadvantage of logistic regression is that most people don't find it intuitive to interpret log odds. 'The probability of being ill is 0.1' is easier to understand than 'the log odds of being ill are –2.197' (compare Table 2.3). However, we can address this problem, because we can transform the log odds back to the scale of a probability.

To see how, consider that we can rewrite the logistic model equation in terms of the probability of the outcome as follows[ii]:

$$p_i = \frac{\exp(\beta_0 + \beta_1 X_{1i} + \beta_2 X_{2i} + \cdots + \beta_k X_{ki})}{1 + \exp(\beta_0 + \beta_1 X_{1i} + \beta_2 X_{2i} + \cdots + \beta_k X_{ki})}$$

It requires a bit of algebra to show that this way of writing the logistic regression model is equivalent to the equation I gave earlier. (I won't show the algebra in this book.) The important thing for us is to note is that we can use logistic regression to predict the probability of an event.

Logistic regression with a single categorical predictor

The simplest type of logistic regression is a model that features just one categorical predictor. Consider again the data in Table 2.1. To analyse these data, we can use logistic regression to model the log odds of classical concert attendance conditional on education. The model equation is as follows:

Model 2.1:

$$\log\left(\frac{p_i}{1 - p_i}\right) = \beta_0 + \beta_1 \times Highschool_i$$

where *Highschool* is a **dummy variable** coded 1 for those who completed high school and 0 for those who did not, and p_i is the probability of attending classical concerts for the *i*th person in our data set. This model proposes that the probability of attending classical concerts depends only on a person's education, and on no other characteristics. That is probably not realistic, but we will explore this simple model as an example. The estimated coefficients are customarily displayed in a table such as Table 2.4. These estimates are found via a mathematical procedure called maximum likelihood estimation, which is explained below in the section 'Logistic Regression: Assumptions and Estimation'.

[ii]If you are not familiar with the *exp*() notation in this equation, please consult Box 2.2.

Table 2.4 Coefficient estimates, standard errors and confidence intervals for the prediction of classical concert attendance by education (Model 2.1)

	Coefficient	SE	95% Confidence Interval
Intercept	−1.48	0.12	
Highschool	1.70	0.17	[1.36, 2.04]

Note. Data were taken from the Family Survey Dutch Population. 95% confidence intervals were computed using the profile likelihood method (see section 'Confidence Intervals'). *SE* = standard error.

So the estimated predictive equation is

$$\log\left(\frac{\hat{p}_i}{1-\hat{p}_i}\right) = -1.48 + 1.70 \times Highschool_i$$

From this equation, we can calculate the predicted log odds of classical concert attendance for those with and without high school education. For those without high school education, *Highschool* = 0, and thus the predictive equation is simply

$$\log\left(\frac{\hat{p}_i \mid \text{No high school}}{1-\hat{p}_i \mid \text{No high school}}\right) = -1.48 + 1.70 \times 0 = -1.48$$

The predicted log odds for those with high school education (*Highschool* = 1):

$$\log\left(\frac{\hat{p}_i \mid \text{High school}}{1-\hat{p}_i \mid \text{High school}}\right) = -1.48 + 1.70 = 0.22$$

Coefficients on the log odds scale, and predicted log odds, are difficult to interpret intuitively. But we can interpret their signs: The negative intercept means that the reference group (here: below high school) has small odds of classical concert attendance.[iii] The positive coefficient for *Highschool* indicates that those with a high school education are more likely to attend classical concerts than those without.

Predicted probabilities

A more intuitive way to understand the predictions from a logistic regression is to calculate predicted probabilities. Using the back-transformation to the probability scale described earlier, we can rewrite our predictive model equation as follows:

[iii]To be precise, negative log odds imply that the odds are below 1, which in turn implies that the probability of the event for the reference groups is below 0.5. Compare Table 2.3 to see why.

$$\hat{p}_i = \frac{\exp(-1.48 + 1.70 \times Highschool_i)}{1 + \exp(-1.48 + 1.70 \times Highschool_i)}$$

Thus, we can calculate the predicted probability of classical concert attendance for the two educational groups:

$$\hat{p}_i \mid \text{No high school} = \frac{\exp(-1.48)}{1 + \exp(-1.48)} = 0.185$$

$$\hat{p}_i \mid \text{High school} = \frac{\exp(-1.48 + 1.70)}{1 + \exp(-1.48 + 1.70)} = 0.555$$

So the model predicts that the probability of attending classical concerts is 0.185 for the less educated and 0.555 for those with high school education or higher.

Estimated odds ratio from a logistic regression model

In many situations, the main aim of the logistic regression model is to estimate the effect of a predictor variable on the odds of the outcome. In our example, the effect of education is expressed in the estimated coefficient of *Highschool*, which is $\hat{\beta} = 1.70$. We can transform this coefficient into an odds ratio by exponentiating it. Thus we calculate

$$\widehat{OR}(\text{High school vs No high school}) = \exp(\hat{\beta}) = \exp(1.70) = 5.48$$

We need to exponentiate the coefficient to calculate the odds ratio, because the coefficient was estimated from a model predicting the logarithm of the odds. **Exponentiation** back-transforms the coefficient to the scale of the odds. If you are not familiar with the concept of exponentiation, or the notation exp(), please consult Box 2.2.

The result of our calculation gives the estimated odds ratio as 5.48. We may interpret this as follows: from this model, we estimate that those with high school education have 5.48 times higher odds of attending classical concerts than those with lower than high school education.

Box 2.2

Exponentiation

Exponentiation is the mathematical operation of raising one number to the power of another. Two numbers are involved: the base b and the power n. We write b^n to denote exponentiating b to the power of n. This means that b is multiplied by itself n times. For example, if $b = 10$ and $n = 2$, then the calculation and its result are as follows: $10^2 = 10 \times 10 = 100$. For any base b, the following are true by definition:

- $b^1 = b$
- $b^0 = 1$
- $b^{-1} = \frac{1}{b}$

The following relationship is a property of exponentiation:

$$b^n = \frac{1}{b^{-n}}$$

So, for example,

$$10^{-2} = \frac{1}{10^2} = 0.01$$

Fractional powers are defined so that $b^{1/n} = \sqrt[n]{b}$. For example,

- $10^{0.5} = \sqrt[2]{10} \approx 3.16$
- $500^{0.1} = \sqrt[10]{500} \approx 1.86$

When the base of an exponentiation is not specified, it is assumed to be Euler's number $e = 2.7182818\ldots$ So if mathematicians or statisticians say 'I exponentiate n', they mean 'I calculate e^n'. An alternative notation for e^n is to write $\exp(n)$. Thus, for example,

- $\exp(2) = e^2 = e \times e = 7.389$
- $\exp(1.7) = e^{1.7} = 5.48$

A logarithm (see Box 2.1) is the inverse function to exponentiation. Thus, for example,

- $\log(7.389) = 2$
- $\log(5.48) = 1.7$

Don't the numbers look familiar?

When considering the predicted probabilities and the estimated odds ratio from the logistic regression model (Model 2.1), you may notice that they are exactly the same as the values we calculated directly from the data (Table 2.1). Indeed, it is the case that a logistic regression model with a single dummy predictor always produces estimated probabilities and odds ratios that match the observed probabilities and odds ratios. This will not in general be the case for models involving a continuous predictor, nor for models with several predictors, as we shall see presently.

Logistic regression with two categorical predictors

Let's now move on to a slightly more complex logistic regression model containing two categorical predictors. Consider Table 2.5, which shows the numbers and percentages of people who attend classical concerts, by education and the presence of young children in the household.

Table 2.5 Classical concert attendance by education level and presence of young children in the household

Young Children in Household?	Classical Concerts	Education					
		Below High School		High School or Higher		Total	
No young kids	Yes	77	20.8%	104	55.6%	181	32.4%
	No	294	79.2%	83	44.4%	377	67.6%
	Total	371	100.0%	187	100.0%	558	100.0%
Young kids	Yes	10	10.1%	33	55.0%	43	27.0%
	No	89	89.9%	27	45.0%	116	73.0%
	Total	99	100.0%	60	100.0%	159	100.0%

Note. Data were from the Family Survey Dutch Population. Young children in the household means that at least one child aged 4 or younger lives in the same household as the respondent.

A simple model with two categorical predictors

We can model how the probability of attending classical concerts varies with education and the presence of young children. For example, we may wish to investigate the hypothesis that higher education is associated with a greater probability of attending classical concerts, and that living with young children is associated with a smaller probability of attending classical concerts, because living with young children may mean that there is less time for such activities. Thus, we might propose the following logistic regression model:

Model 2.2:

$$\log\left(\frac{p_i}{1-p_i}\right) = \beta_0 + \beta_1 Highschool_i + \beta_2 Youngkids_i$$

where

- $\hat{p}_i$ is the predicted probability of attending classical concerts for the ith person, as before;
- *Highschool* is a dummy variable coded 1 for those with high school education, and 0 for others; and

- *Youngkids* is a dummy variable coded 1 if at least one child aged 4 or younger lives in the same household as the respondent, and 0 otherwise.

Using Model 2.2, we could estimate the effect of *Highschool* on classical concert attendance, controlling for the presence or absence of young children. We can also estimate the effect of *Youngkids* on classical concert attendance, controlling for education. The principle of statistically controlling for another predictor variable is applied in logistic regression in the same way as in linear regression (see *The SAGE Quantitative Research Kit*, Volume 7).

Modelling an interaction

Model 2.2 makes the assumption that the effect of having young kids in the household is the same among both educational groups. This is because the model does not feature an interaction between the two variables. Generally speaking, a model without an interaction makes the assumption that the effect of one predictor does not depend on the value of another.

However, sometimes we have reason to think that the effect of one predictor might depend on the value of another predictor. For example, we might consider that those with high school education might tend to have the money to pay for child care, and therefore might be able to go to classical concerts even if they have young children, while those with lower education tend to have less money and might need to sacrifice their interest in classical concerts (if they have such an interest) to look after their children instead. So the presence of young children would affect those with high school education differently to those without high school education. This is an example of an interaction (here: a hypothesised interaction).[iv] If we thought that such an interaction might play a role, we might specify a model such as Model 2.3.

Model 2.3:

$$\log\left(\frac{p_i}{1 - p_i}\right) = \beta_0 + \beta_1 Highschool_i + \beta_2 Youngkids_i + \beta_3 Highschool_i \times Youngkids_i$$

The difference between the two models is that Model 2.3 allows for an interaction between *Highschool* and *Youngkids*, while Model 2.2 does not. Both models assume that the probability of attending classical concerts depends only on a person's education

---

[iv]You may have noticed that, in explaining the interaction, I have taken recourse to a third variable, namely money, or income. Strictly speaking, if I thought my explanation was true, I should also include income as a predictor variable in my model. I won't do so in this section, because I want to keep the examples simple to begin with. We will look at logistic regression models with more than two predictors later on.

and the presence or absence of young children. Model 2.2 additionally assumes that the relationship between each predictor and the outcome is the same for each level of the other predictor. Model 2.3, in contrast, allows the relationship between each predictor and the outcome to vary depending on the level of the other predictor.

Illustrating the models with and without an interaction

Table 2.6 shows the results from estimating Models 2.2 and 2.3 on the data displayed in Table 2.5.

Table 2.6 Coefficient estimates, standard errors and confidence intervals for the prediction of classical concert attendance by education and presence of young children in the household (Models 2.2 and 2.3)

	Model 2.2: No Interaction			Model 2.3: With Interaction		
	Coefficient	SE	OR	Coefficient	SE	OR
Intercept	−1.41	0.12		−1.34	0.13	
Highschool	1.72	0.18	5.60	1.57	0.20	4.78
Youngkids	−0.39	0.22	0.68	−0.85	0.36	0.43
Highschool × Youngkids				0.82	0.46	2.27

Note. Data were taken from the Family Survey Dutch Population. SE = Standard error; OR = odds ratio.

The estimated logistic regression equation for Model 2.2 is:

$$\ln\left(\frac{\hat{p}_i}{1-\hat{p}_i}\right) = -1.41 + 1.72 \times Highschool_i - 0.39 \times Youngkids_i$$

The estimated logistic regression equation for Model 2.3 is:

$$\ln\left(\frac{\hat{p}_i}{1-\hat{p}_i}\right) = -1.34 + 1.57 \times Highschool_i - 0.85 \times Youngkids_i + 0.82 \times Highschool_i \times Youngkids_i$$

Both models feature more than one predictor, so just like in multiple linear regression, we need to take account of this in our interpretation of coefficients. In Model 2.2, the slope coefficients are to be interpreted as the difference in the log odds of the outcome associated with a 1-point change in the predictor, *holding all other predictor variables constant*. Thus, Model 2.2 estimates that:

- Having a high school education is associated with log odds of attending classical concerts that are higher by 1.72 compared to having a lower educational level, controlling for the presence or absence of young children in the household.

- Having young children is associated with a reduction of 0.39 in the log odds of attending classical concerts compared to no children, keeping educational level constant.

In interpreting the coefficients of Model 2.3, we additionally need to take care to interpret all coefficients in the light of the interaction effect. The reference group here is those with lower than high school education (*Highschool* = 0) and without young children in the household (*Youngkids* = 0). Thus,

- Among those without young children, having a high school education is associated with 1.57 higher log odds of attending classical concerts.
- Among those with lower than high school education, having young children in the household is associated with 0.85 lower log odds of attending classical concerts.
- Among those with young children, having a high school education is associated with 1.57 + 0.82 = 2.39 higher log odds of attending classical concerts.
- Among those with high school education, the presence or absence of young children is estimated to have almost no relationship with the log odds of attending classical concerts, since –0.85 + 0.82 = –0.03 ≈ 0.

Odds ratios

Model 2.2 estimates the odds ratios for *Highschool* and *Youngkids* to be

$$\widehat{OR}_{Highschool} = \exp(1.72) = 5.60$$

$$\widehat{OR}_{Youngkids} = \exp(-0.39) = 0.68$$

So Model 2.2 estimates that:

- High school-educated people have 5.6 times higher odds of attending classical concerts than less well educated people, controlling for the presence of young children in the household.
- Respondents with young children in the household have about 32% lower odds of attending classical concerts than respondents without young children.

In Model 2.3, the interaction means that interpretation is a bit more complex. According to the results of Model 2.3,

- Among those with young children, a high school education is associated with 4.78 times higher odds of attending classical concerts.
- Among those without young children, a high school education is associated with 4.78 × 2.42 = 10.9 times higher odds of attending classical concerts. Note that on the odds ratio scale, effects of interactions are multiplicative.

- Among those with lower than high school education, having young children is associated with 57% lower odds (1 − 0.43 = 57%) of attending classical concerts.
- Among those with a high school education, having young children does not appear to affect the odds of attending classical concerts very much (since 0.43 × 2.27 = 0.98 ≈ 1).

So, according to the estimates from Model 2.2, people with high school education are more likely to attend classical concerts than those with lower education, and people with young children are less likely to attend classical concerts than people without young children. According to Model 2.3, the presence of young children affects the probability of classical concert attendance only among those with below high school education, whereas young children have barely any effect on concert attendance among the high school educated.

Which of these two models is more consistent with the data? We won't decide that question for now, but we will return to it in the section 'Logistic Regression: Assumptions and Estimation', when we consider formal model comparisons.

Predicted probabilities

From the estimates of Models 2.2 and 2.3, we can calculate predicted probabilities of attending classical concerts (Table 2.7). Of course, the predicted probabilities for the different groups will not be the same for the two models. The formula for predicting probabilities from a logistic regression model was shown above in the section 'Log Odds: The Logit Transformation'.

Table 2.7 Predicted probabilities of attending classical concerts from two logistic regression models (2.2 and 2.3)

Predictor		Predicted Probability of Attending Classical Concerts	
Highschool	Youngkids	Model 2.2 (No Interaction)	Model 2.3 (With Interaction)
0	0	0.197	0.208
0	1	0.142	0.101
1	0	0.578	0.556
1	1	0.482	0.550

Note. Data were taken from the Family Survey Dutch Population.

The predicted probabilities offer a different way for us to understand the model results and how the two models (2.2 and 2.3) imply quite different conclusions. For example, according to Model 2.2, there is a considerable difference in the probability of classical concert attendance among people with high school education, depending on whether they have young children or not (0.482 when young children are present vs 0.578 when there are none). In contrast, according to Model 2.3, among those with high school education the presence of young children hardly affects the probability of classical concert attendance at all (0.550 vs 0.556).

Another thing that you may have noticed is that the predicted probabilities from Model 2.3 exactly match the observed proportions reported in Table 2.5. In general, when all predictors in a logistic regression are categorical, the following is true: if the model contains all possible interactions between the categorical predictors, then the predicted probabilities will exactly match the observed proportions. Note, however, that for more than two predictors, 'all possible interactions' means 'the two-way interactions between all possible pairs of predictors as well as all possible higher order interactions (three-way interactions, four-way interactions and so forth)'. If the model contains no interactions, if it contains only some of the possible interactions, or if it contains numeric predictors, then in general the predicted probabilities won't match the observed proportions exactly.

Logistic regression with a numeric predictor

Let's now consider logistic regression with a numeric predictor. As an example, suppose that we are interested in estimating the relationship between the probability of attending classical concerts and age. Figure 2.1 shows the distribution of age in the data from the FSDP.

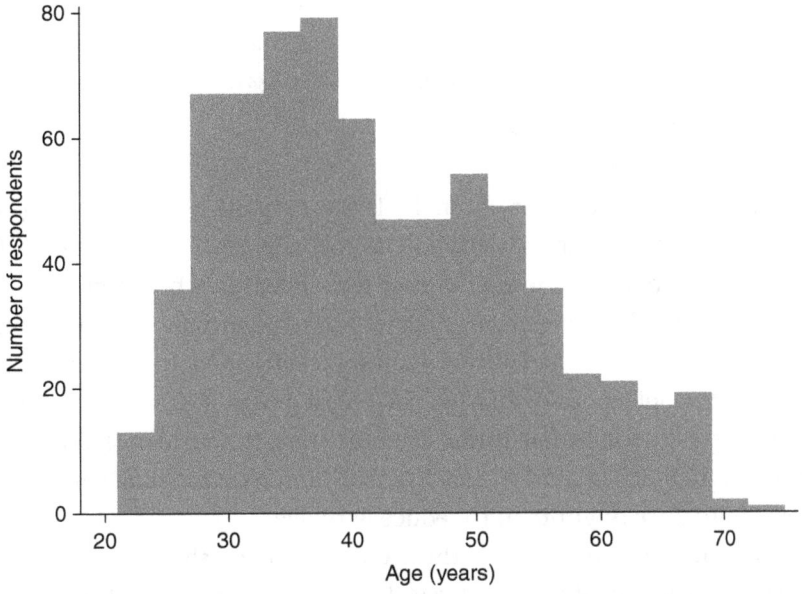

Figure 2.1 Age distribution of respondents in the Family Survey Dutch Population

Note. Data were taken from the Family Survey Dutch Population.

What does the relationship between age and classical concert attendance look like in this data set? To explore the relationship between a continuous predictor and a binary outcome, it is useful to group the respondents according to their values on the continuous predictor, enabling us to construct tables and plots that show how the outcome varies with the grouped version of the predictor variable.

Consider Table 2.8. I have grouped age into 10 decile groups. The first decile group contains approximately the 10% youngest respondents, the last group contains approximately the 10% oldest respondents. (Because age is measured in discrete years in this data set, the groups don't end up being exactly the same size.) Looking at Table 2.8, we observe that, for the most part, the proportion of classical concert goers appears to increase with age, although this is not a monotonous increase in this sample.

Table 2.8 The relationship between age and attending classical concerts

Age Decile Group	Number of Respondents	Age Range	Mean Age in This Group	Percent Attending Classical Concerts
1	92	21–29	26.8	20.7%
2	70	30–32	30.9	24.3%
3	74	33–35	34.0	27.0%
4	51	36–37	36.5	33.3%
5	77	38–40	38.9	28.6%
6	68	41–44	42.4	20.6%
7	78	45–49	46.9	33.3%
8	79	50–53	51.5	41.8%
9	59	54–59	56.1	47.5%
10	69	60–75	64.6	40.6%

Note. Data were taken from the Family Survey Dutch Population.

Figure 2.2 illustrates the results from Table 2.8 by plotting the mean age in each group against the percentage that attend classical concerts. This is a kind of scatter plot, but the points here represent groups of respondents, rather than individuals.

Figure 2.2 suggests that, in general, the older respondents are, the more likely they are to attend classical concerts. There are some exceptions to this pattern in this data set for groups of people around 40. Also, the oldest group of respondents (those aged 60 and older) report somewhat lower classical concert attendance than groups of respondents in their 50s, maybe suggesting that from retirement age onwards, classical concert attendance might tend to reduce with age.

Overall, it is not easy to say, from these data, what the shape of the relationship between age and the probability of classical concert attendance is. For example, it is

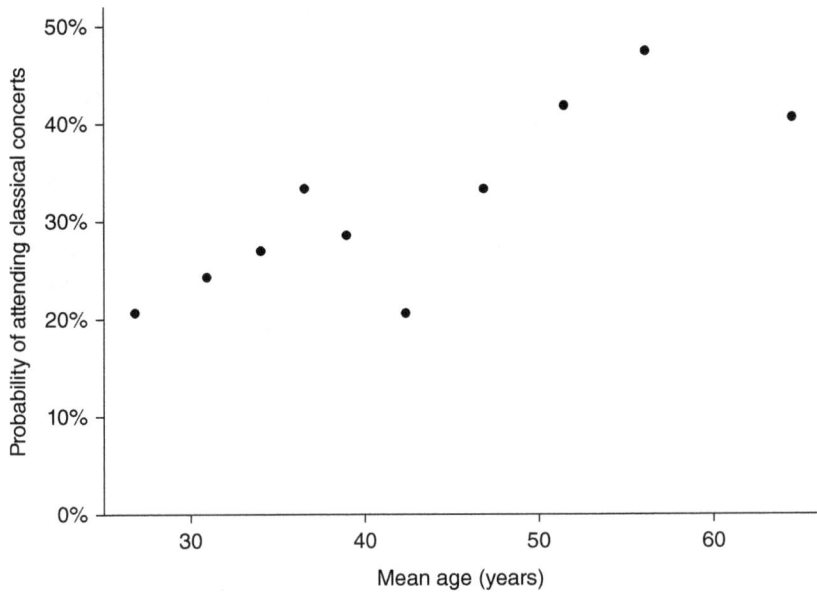

Figure 2.2 The relationship between age and percentage attending classical concerts

Note. This figure plots the percentage attending classical concerts in each of 10 groups defined by the deciles of age. The position of each dot in the *x*-axis direction denotes the mean age in the group. Data were taken from the Family Survey Dutch Population.

unclear whether this relationship can be represented by a straight line or not. In the section 'Interpretation of Effect Sizes and Graphical Illustration' below, we will consider in detail how to explore the shape of a relationship between a continuous predictor and a binary outcome probability. For now, let's assume a simple model for this relationship to enable us to focus on understanding how logistic regression with a numeric predictor works.

Let's propose the following model:

Model 2.4:

$$\ln\left(\frac{p_i}{1 - p_i}\right) = \beta_0 + \beta_1 AgeC_i$$

where $AgeC_i$ is the age (in years) of respondent i, centred at 40 years, which happens to be approximately the mean age in this sample. 'Centred at 40 years' means that we have calculated the variable $AgeC$ as $AgeC = Age - 40$. It is not strictly necessary to centre age, but it has the advantage that those with average age (40 years) have the value 0 on our predictor variable. This makes the intercept interpretable as the log odds of those of average age.[v]

Although we've grouped the respondents for the purpose of data exploration (in Table 2.8 and Figure 2.2), Model 2.4 considers age as a continuous variable and takes into account each respondent's age individually. Model 2.4 assumes that the log odds of attending classical concerts are associated with age in a linear way. The model also assumes that no other variables play a part in the prediction of classical concert attendance, once age has been accounted for. (The latter assumption is obviously not realistic; we will look at a model with more predictors later.) Estimates from fitting Model 2.4 on the FSDP data are shown in Table 2.9.

Table 2.9 Estimates from a logistic regression predicting classical concert attendance by age (Model 2.4)

	Coefficient	SE	Odds Ratio	95% Confidence Interval for Odds Ratio
Intercept	−0.88	0.09		
Age (centred at 40)	0.029	0.007	1.029	[1.015, 1.043]

Note. 95% confidence intervals were computed using the profile likelihood method. *SE* = standard error.

Since *AgeC* is a continuous variable measured in years, the coefficient represents the average difference in the log odds of attending classical concerts associated with a 1-year age difference. The estimated logistic regression equation is as follows:

$$\ln\left(\frac{\hat{p}_i}{1-\hat{p}_i}\right) = -0.88 + 0.029 \times AgeC_i$$

The estimated odds ratio is calculated thus:

$$\widehat{OR} = \exp(0.029) = 1.029$$

This is to be interpreted as follows: the model predicts that with every age year, the odds of attending classical concerts increase by a factor of 1.029, that is, by 2.9%.

Model 2.4 specifies that the relationship between age and classical concert attendance is additive on the log odds scale. This implies that it is multiplicative on the scale of the odds (see Box 2.1). For example, if we wish to know the difference in the

---

[v] It is perfectly all right to use non-centred age as a predictor in our model. If we did this, our estimate of the slope coefficient of age would have been exactly the same as for *AgeC*, but the intercept would be a different number and have a different interpretation. Without **centring**, the intercept would indicate the predicted log odds of concert attendance for a person aged 0, which of course is not very meaningful.

odds of classical concert attendance associated with a difference of 10 years (rather than 1 year), we need to calculate the following:

$$\widehat{OR}_{10years} = \widehat{OR}_{1year}^{10} = 1.029^{10} \approx 1.34$$

Alternatively, we can get the same result as follows:

$$\widehat{OR}_{10years} = \exp(0.029 \times 10) = \exp(0.29) \approx 1.34$$

So a 10-year difference in age is associated with about 34% increased odds of attending classical concerts, according to the estimate from Model 2.4.

We might also be interested in transforming our results back onto the probability scale. The predicted probability of attending classical concerts is estimated to be

$$\hat{p}_i = \frac{\exp(-0.88 + 0.029 \times AgeC_i)}{1 + \exp(-0.88 + 0.029 \times AgeC_i)}$$

We can use this equation to calculate the predicted probability of attending classical concerts given any particular age. For example, for a person aged 40, $AgeC = 0$, and the predicted probability of attending classical concerts is

$$\hat{p}_i \mid (AgeC = 0) = \frac{\exp(-0.88 + 0.029 \times 0)}{1 + \exp(-0.88 + 0.029 \times 0)} = \frac{\exp(-0.88)}{1 + \exp(-0.88)}$$
$$= 0.29$$

A formal way to interpret this is as follows: the model predicts that a 40-year-old respondent has a 29% probability of attending classical concerts. An equally valid formulation in this case is that the model estimates that 29% of 40-year-olds in the Dutch population attend classical concerts. Of course, we could do an analogous calculation for any value of age. For example, a 60-year-old person has $AgeC = 20$, and their predicted probability of classical concert attendance is calculated as follows:

$$\hat{p}_i \mid (AgeC = 20) = \frac{\exp(-0.88 + 0.029 \times 20)}{1 + \exp(-0.88 + 0.029 \times 20)} = \frac{\exp(-0.3)}{1 + \exp(-0.3)}$$
$$= 0.43$$

So, according to Model 2.4, a 60-year-old person has a 43% probability to attend classical concerts – considerably higher than a 40-year-old.

To further understand the modelled effect of age on our outcome, and to investigate how well the model fits the data, we might visualise the predicted probabilities in a graph such as Figure 2.3. This shows the model-predicted probabilities as a function of age. Notice that the line is not exactly straight, but slightly curved: although our prediction is linear on the log odds scale, it is not linear on the probability scale (see also section 'The logistic curve' above).

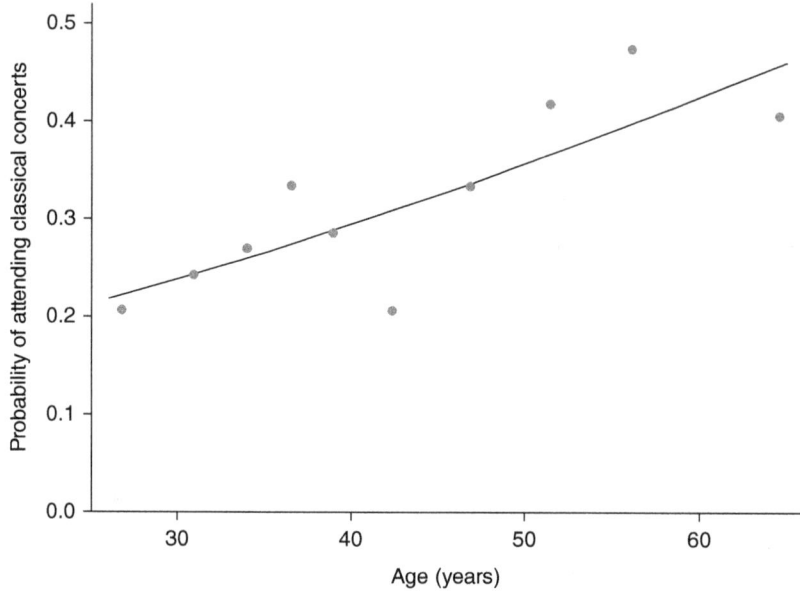

Figure 2.3 Observed proportions and predicted probability of classical concert attendance by age decile (Model 2.4)

Note. The black line represents the predicted probabilities from Model 2.4. The grey dots show the 10 decile groups of the age distribution, plotting the mean age of each group on the horizontal axis, and on the vertical axis the observed proportion that report attending classical concerts.

We can use Figure 2.3 to evaluate whether the shape of the relationship between our predictor and the probability of the outcome follows the shape implied by the logistic regression model. In a well-fitting model, all the observed proportions would be very close to the predicted probabilities: the dots would all lie close to the line. In our case, in a few groups the predictions differ considerably from the observed proportions, so the model does not fit perfectly. On the other hand, Figure 2.3 does not reveal an obvious plausible alternative pattern, so it is not clear whether we can improve the model by transforming our age variable in some way. In this case, the most likely reason for the relatively poor fit is that our model is too simple: it fails to contain several important predictors of classical concert attendance. Thus, the most important way in which we could improve model fit is likely to be the addition of relevant additional predictors. We will look at assessing model fit in detail in the section 'Model Comparison and Hypothesis Tests'.

Logistic regression: assumptions and estimation

This section will look at logistic regression from a more formal point of view, which is necessary for us to understand the assumptions that underpin the model.

The binomial distribution

Logistic regression assumes that the outcome variable follows a *binomial distribution*, conditional on the predictor variables.[vi] The simplest example of a binomial distribution results from tossing a coin. A coin has two sides, traditionally called 'heads' and 'tails'. So the outcome of a toin coss is a dichotomous variable. Let's assume that the two events – 'heads' or 'tails' – are equally likely. This is what is called a 'fair coin'. With a fair coin, the probability of a single coin toss resulting in the outcome 'heads' is 0.5. Now, imagine a game where I toss the coin four times and count the number of heads. The possible outcome events of this game are the numbers from 0 to 4. So there are five possible outcome events, which differ in how probable they are. For example, it is very unlikely that I get 'heads' four times in a row. The binomial distribution allows me to determine exactly how likely each event is. Box 2.3 explains the binomial distribution more formally and in more detail.

Box 2.3

The Binomial Distribution

The binomial distribution describes the probability (p) of getting a number of events y in a binary outcome, given a sample size (n) and the probability of the event (π). To give a simple example, imagine we toss a fair coin four times and count the number of heads. Then $\pi = 0.5$, $n = 4$, and we want to know the probability $p(y)$ of getting heads $y = 0, 1, 2, 3,$ or 4 times.

This probability can be calculated by a formula called the probability mass function of the binomial distribution:

$$p(y) = \frac{n!}{y!(n-y)!} \pi^y (1-\pi)^{n-y}$$

(Continued)

[vi]In linear regression, the assumption is that the outcome follows a normal distribution, conditional on the predictor variables. This is the same as saying that the errors from a linear regression model are assumed to follow a normal distribution.

This equation allows us to calculate the probability, $p(y)$, of observing y events, under the assumption that our outcome follows a binomial distribution with event probability π. The exclamation mark symbol (!) indicates a factorial, which is defined as follows: $y! = 1 \times 2 \times 3 \times \cdots \ldots \times y$. For example, if $y = 3$, then $y! = 1 \times 2 \times 3 = 6$. Also, note that $0! = 1$ by definition.

For our example with $\pi = 0.5$ and $n = 4$, the formula gives

$$p(y) = \frac{4!}{y!(4-y)!} 0.5^y (1-0.5)^{4-y}$$
$$= \frac{4!}{y!(4-y)!} 0.5^4$$

From this, we can calculate the probability of obtaining 0, 1, 2, 3, or 4 heads as in Table 2.10.

Table 2.10 Example calculations for a binomial distribution with $\pi = 0.5$ and $n = 4$

y	$p(y) = \dfrac{4!}{y!(4-y)!} 0.5^4$
0	$\dfrac{4!}{0!(4-0)!} 0.5^4 = 1 \times 0.5^4 = 0.0625$
1	$\dfrac{4!}{1!(4-1)!} 0.5^4 = 4 \times 0.5^4 = 0.25$
2	$\dfrac{4!}{2!(4-2)!} 0.5^4 = 6 \times 0.5^4 = 0.375$
3	$\dfrac{4!}{3!(4-3)!} 0.5^4 = 4 \times 0.5^4 = 0.25$
4	$\dfrac{4!}{0!(4-0)!} 0.5^4 = 1 \times 0.5^4 = 0.0625$

So for a binomial distribution with $n = 4$ and $\pi = 0.5$, the probability of observing 0 events is 0.0625, the probability of observing 1 event is 0.25, and so forth.

In a fair coin toss, the probability of the event 'heads' is $\pi = 0.5$, but the binomial distribution can be calculated for any event probability between 0 and 1. The shape of the binomial distribution varies with the event probability: if $\pi < 0.5$, the distribution is positively skewed; if $\pi = 0.5$, it is symmetric; and if $\pi > 0.5$, it is negatively skewed. See Figure 2.4 for an illustration of binomial distributions, with $n = 30$ and three different values of π. For large sample sizes and π close to 0.5, the binomial distribution looks very similar to the normal distribution.

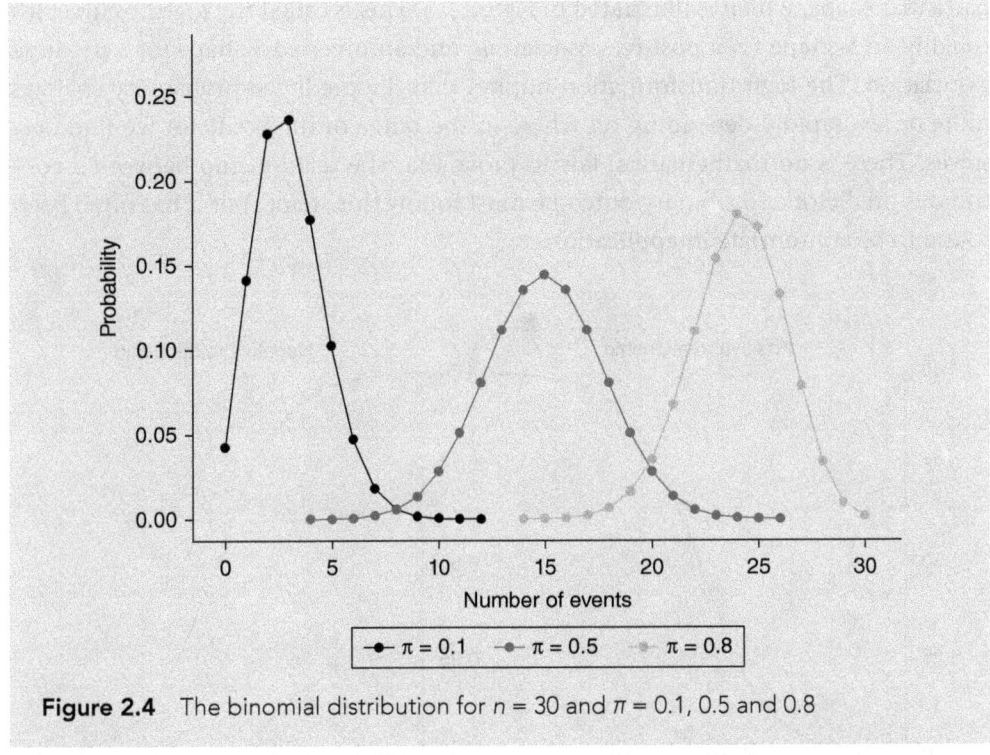

Figure 2.4 The binomial distribution for n = 30 and π = 0.1, 0.5 and 0.8

In logistic regression, we assume that the number of outcome event follows a binomial distribution with event probability π, where π is conditional on the values of the predictors. For example, our results of Model 2.4 suggested that a randomly chosen 40-year-old Dutch person has a probability of $\pi = 0.29$ to be someone who attends classical concerts. If the model is correct, then if we repeatedly take samples of size n from the Dutch population of 40-year-olds, we would find that the number of classical concert goers across these samples follows a binomial distribution with $\pi = 0.29$. One way in which this assumption can be violated is when we have failed to include an important predictor in our model. In such a case, the sampling distribution of the proportion of concert goers may not be binomial, but instead, for example, may exhibit more variation than predicted by the binomial distribution. In that case, confidence intervals and statistical tests about model coefficients may not be accurate.

The logistic curve

Another important assumption comes into play for all logistic regression models with at least one continuous predictor. The logit transformation implies that the relationship between a continuous predictor and the probability of an outcome follows a particular shape, which is illustrated in Figure 2.5. This is called the logistic curve: it's roughly an S-shape for a positive association, and an inverted S-shape for a negative association. The logit transformation implies that the predicted probability changes more or less rapidly depending on where in the range of the predictor we find ourselves. There is no mathematical law to prove that the relationship between a continuous predictor and a binary outcome must follow this shape, but it has often been found to be appropriate in applications.

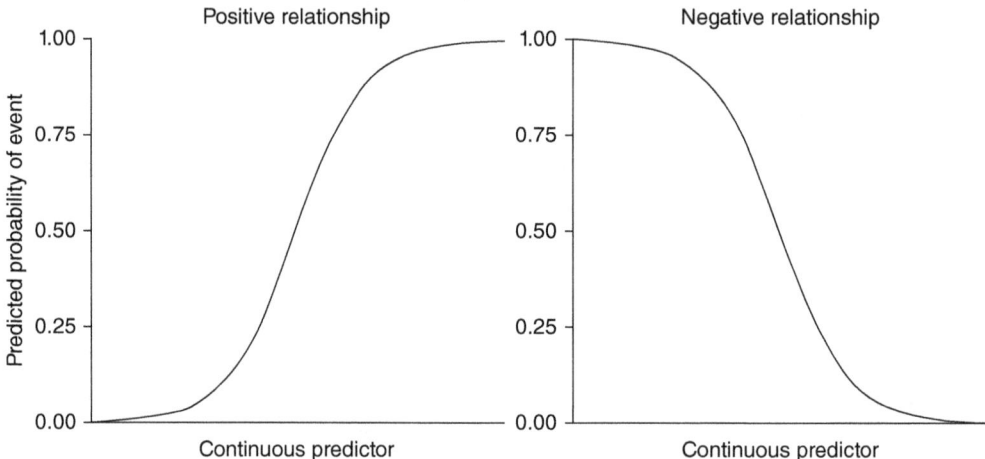

Figure 2.5 The shape of the relationship between a numeric predictor and the probability of the outcome event, as assumed by the logistic regression model

In practice, the logistic shape may not be fully realised in a particular data set, because the continuous predictor may be restricted in its range. Consider for example Model 2.4, which proposed age as the only predictor of classical concert attendance. If you leaf back to Figure 2.3, which illustrates how the predicted probabilities vary with age, you might ask yourself why this does not show the S-shape I just told you was always implied by the logistic model. The reason is that I only plotted predicted probabilities for ages between 21 and 75, which represents the age range of our sample. Figure 2.6 shows the full relationship between age and the predicted probability of classical concert attendance implied by the model. Here, we can see the S-shape. Thus, mathematically Model 2.4 does posit an S-shaped relationship between age and the probability of concert attendance. From a practical point of view, however, we should not extrapolate predictions of the probability of classical concert attendance for values of age outside the range in the data set that we used to estimate the model.

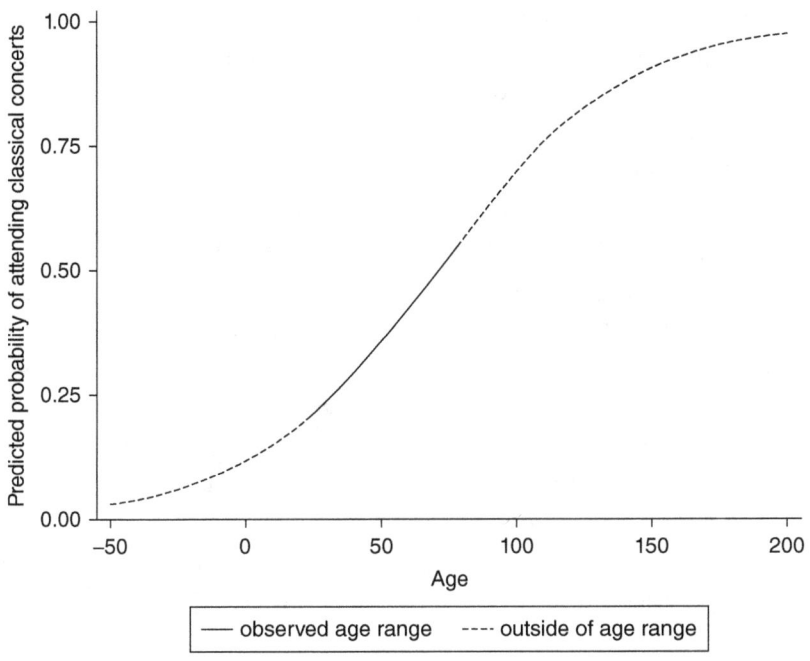

Figure 2.6 The relationship between age and classical concert attendance predicted by Model 2.4 (mathematical prediction beyond age range in the Family Survey Dutch Population data set)

Maximum likelihood estimation: how the coefficients are found

How are the coefficients in a logistic regression estimated from a data set? You may recall that we find the coefficient estimates for a linear regression by the method of least squares (see *The SAGE Quantitative Research Kit*, Volume 7). For logistic regression, a different procedure is used, which is called maximum likelihood estimation.

The mathematics of **maximum likelihood estimation** are beyond the scope of this book.[vii] But we can understand the procedure conceptually. First, consider the concept of a likelihood: this is a measure of how likely the observed data are given a set of coefficients from a model. For example, consider again Model 2.3:

Model 2.3:

$$\ln\left(\frac{p_i}{1-p_i}\right) = \beta_0 + \beta_1 Highschool_i + \beta_2 Youngkids_i + \beta_3 Highschool_i \times Youngkids_i$$

[vii]If you are interested, have a look at Hilbe (2009, Chapter 3).

We wish to estimate four coefficients: β_0, β_1, β_2, β_3. Imagine now that we could look at a particular combination of values for these four coefficients, picking four arbitrary numbers. Then it is possible to calculate how likely we were to observe the data we did observe (our data set) if our four arbitrary numbers were the true coefficients. The result of this calculation is called the likelihood.[viii] The idea of maximum likelihood estimation is to find the set of coefficients that produces the largest likelihood. This is called 'maximising the likelihood function'. For example, in Model 2.3, the set of coefficients reported in Table 2.6 is that which produces the largest likelihood. Any other set of coefficients would have produced a smaller likelihood.

The likelihood is usually a very small number. For example, the likelihood for Model 2.3 is approximately 2.9×10^{-188} – that is, 0.0000 . . . 00029, with altogether 188 zeroes. The value of the likelihood in itself is not very meaningful, but its natural logarithm, called the log likelihood (or LL for short), is often reported by researchers writing up the results from a logistic regression, and will be important for statistical hypothesis tests, as we shall see below. The log likelihood of Model 2.3 is LL = −431.8.

Confidence intervals

A typical results table for a logistic regression includes the estimated standard errors and confidence intervals for the coefficients (see e.g. Table 2.9 or Table 2.10). The standard errors are a measure of the variability of our coefficient estimates between samples. The larger the standard error, the less precise is our estimate. It is beyond the scope of this book to explain exactly how the standard errors are calculated, but if you are interested, have a look at Hilbe (2009, p. 102f). Many software applications display, by default, confidence intervals for coefficients of a logistic regression in a fashion very similar to a linear regression, by calculating, for example, the 95% CI as follows:

$$CI_{\beta,0.95} = \hat{\beta} \pm z_{0.975} \times SE_{\hat{\beta}}$$

[viii]In order to calculate the likelihood, we need to make certain assumptions. For example, to calculate the likelihood for a logistic regression, we assume that the proportion of people with $Y = 1$ follows a binomial distribution for any given combination of predictor values. In other types of models, which also use maximum likelihood estimation, we might make different distributional assumptions. For example, in Poisson regression, which is discussed in Chapter 5 of this book, the likelihood is calculated under the assumption that the outcome variable follows a Poisson distribution, conditional on the predictors.

where $SE_{\hat{\beta}}$ is the estimated standard error for the coefficient estimate $\hat{\beta}$, and $z_{0.975}$ is the 97.5th percentile of the normal distribution, which is approximately 1.96.[ix] Confidence intervals calculated in this way are called *Wald confidence intervals*. For Wald confidence intervals to be correct, we need to assume that the coefficient estimates have a sampling distribution that is approximately normal. This is called the *normal approximation*. It is a realistic assumption for large samples, and when the overall rate of events in our population is not very close to 0 or 1.

A different type of confidence interval, which is accurate even when samples are small and the rate of events is close to 0 or close to 1, is called the *profile likelihood confidence interval* (see e.g. Agresti, 2013, p. 80). Profile likelihood confidence intervals are calculated from the likelihood function, and good software programmes will allow you to obtain them. The intervals displayed in Table 2.11 are profile likelihood confidence intervals. In many cases, the two sets of confidence intervals (based on normal approximation and based on profile likelihood) will be very similar. For example, according to our estimates from Model 2.3, the Wald 95% CI for the coefficient for *Highschool* is almost the same as the profile likelihood CI (Wald CI [1.18, 1.95]; profile likelihood CI [1.19, 1.95]. The confidence intervals for *Youngkids* differ a bit more, but the differences are still very small (Wald CI [−1.55, −0.15]; profile likelihood CI [−1.60, −0.19]). If you do find an important difference between the two methods, I recommend trusting the profile likelihood CIs, and in this chapter, I will generally report profile likelihood CIs. Note, however, that most publications by applied researchers in the social sciences will report Wald CIs, largely because Wald CIs are the default in most software packages.

Table 2.11 Estimated coefficients and odds ratio from Model 2.3

	Log Odds Scale			OR Scale	
	Coef.	SE	95% CI for Coef.	OR	95% CI for OR
Intercept	−1.34	0.13	[−1.60, −1.09]		
Highschool	1.57	0.20	[1.19, 1.95]	4.78	[3.27, 7.03]
Youngkids	−0.85	0.36	[−1.60, −0.19]	0.43	[0.20, 0.83]
Highschool × Youngkids	0.82	0.47	[−0.07, 1.77]	2.27	[0.94, 5.86]

Note. Data were taken from the Family Survey Dutch Population. 95% confidence intervals were calculated using the profile likelihood method. Coef. = coefficient; *SE* = standard error; CI = confidence interval; OR = odds ratio.

[ix]For a 99% CI, we would have needed to use, the 99.5th percentile of the normal distribution, which is approximately 2.58.

We can calculate a confidence interval for the odds ratio as well. The confidence limits around the odds ratio are derived from the confidence limits on the log odds scale by exponentiation. For example, consider Table 2.11. The lower limit for the odds ratio of *Highschool* is calculated as $L = \exp(1.19) = 3.27$. Similarly, the upper limit is $U = \exp(1.95) = 7.03$.

Note that the confidence interval for the odds ratio is not symmetric around the point estimate of the odds ratio. That is, the confidence interval's upper limit does not have the same distance from the point estimate of the odds ratio as the lower limit. This is generally so. On the log odds scale, Wald confidence intervals are symmetric around the point estimate by definition, but profile likelihood confidence intervals may be symmetric or asymmetric on the log odds scale.

Model comparison and hypothesis tests

Let's now consider methods for formally comparing models, with the purpose of evaluating which of two models provides a better fit to the data. Consider two nested models,[x] a smaller Model A and a larger Model B. If both models were linear regression models, we could use the *F*-test for model comparison to assess which model we might prefer (see *The SAGE Quantitative Research Kit*, Volume 7).

However, if Models A and B are logistic regression models, we cannot apply the *F*-test, since the test relies on the assumption that errors are normally distributed and homoscedastic. Errors from a logistic model are never homoscedastic and are unlikely to follow a normal distribution. So in logistic regression, instead of the *F*-test, we can use a different test for model comparisons. This is the **likelihood ratio test** (LRT), and it is based on the log likelihood of the models (see the section 'Maximum Likelihood Estimation').

Likelihood ratio test

We can formally compare two nested logistic regression models using the likelihood ratio test. The test statistic is

$$\Lambda = -2 \times \left(LL_{smaller\ model} - LL_{larger\ model} \right)$$

where Λ (the Greek letter capital lambda) denotes the test statistic, and LL signifies the log likelihood (see the section 'Maximum Likelihood Estimation'). So the test statistic is calculated as minus twice the difference between the log likelihoods of the two models we wish to compare.

---

[x]For a definition of nested models, see *The SAGE Quantitative Research Kit*, Volume 7.

The null hypothesis of the LRT is that there is no difference between the two models in how well they predict the outcome. The alternative hypothesis is that the larger model provides a better prediction than the smaller model. If all model assumptions are satisfied, and if the null hypothesis is true, Λ follows a χ^2-distribution with degrees of freedom equal to the difference in the number of **parameters** between the two models.

For example, let's suppose we wish to predict classical concert attendance via education and the presence of children in the house. We then wish to compare Model 2.2 (no interaction between the predictors) with Model 2.3 (with interaction). So we have (see the section 'A Simple Model with Two Categorical Predictors'):

Model 2.2:

$$\log\left(\frac{p_i}{1 - p_i}\right) = \beta_0 + \beta_1 Highschool_i + \beta_2 Youngkids_i$$

Model 2.3:

$$\log\left(\frac{p_i}{1 - p_i}\right) = \beta_0 + \beta_1 Highschool_i + \beta_2 Youngkids_i + \beta_3 Highschool_i \times Youngkids_i$$

These two models are nested, since we can turn Model 2.3 into Model 2.2 by setting the coefficient for the interaction effect to zero (i.e. setting $\beta_3 = 0$). The null hypothesis of the LRT is that the two models predict classical concert attendance equally well. In our example, this is equivalent to hypothesising that there is no interaction between *Highschool* and *Youngkids* (i.e. we hypothesise that $\beta_3 = 0$). The alternative hypothesis is that Model 2.3 does better than Model 2.2 in predicting classical concert attendance. In our example, this is equivalent to hypothesising that there is an interaction between *Highschool* and *Youngkids*, which would imply that $\beta_3 \neq 0$.

The results of an LRT are often displayed in a table such as Table 2.12. The number of parameters of each model is simply the number of its coefficients (slopes and intercept) that are being estimated. The degrees of freedom (*df*) for each model are equal to the sample size minus the number of parameters.

Table 2.12 Likelihood ratio test of Model 2.3 versus Model 2.2

Model		LL	Λ	Parameters	df	df diff.	p
Model 2.2	No interaction	−393.24		3	714		
Model 2.3	With interaction	−391.59	3.28	4	713	1	0.070

Note. LL = log likelihood; *df* = degrees of freedom; *df* diff. = difference in degrees of freedom between the two models.

A brief way of reporting the results of Table 2.12 is as follows: $\Lambda = 3.28$, $df = 1$, $p = 0.070$. As in any statistical hypothesis test, the p-value of an LRT is an indication of the strength of the statistical evidence against the null hypothesis from the data set at hand. A small p-value indicates evidence against the smaller model. A small p-value does not necessarily indicate that the larger model is 'right', but it does indicate evidence that the larger model leads to better predictions than the smaller one. However, there may well be other models (outside of those we compared in our test) that might provide even better predictions.

Conversely, if the LRT results in a large p-value, this indicates that there is little evidence that the larger model improves the prediction of the outcome relative to the smaller model. A large p-value does not necessarily indicate that the smaller model is 'right'. There may well be other models that fit the data better.

In this case, our p-value is 0.07, suggesting that there is moderate evidence against the null hypothesis. Since this value is larger than 0.05, researchers who use the conventional significance threshold would interpret this to say that there is insufficient evidence to conclude that Model 2.3 improves the prediction of classical concert attendance relative to Model 2.2, or in other words, that there is insufficient evidence to support the idea that the interaction *Highschool × Youngkids* improves the prediction of our outcome. On the other hand, if the p-value had been smaller than 0.05, researchers who use this threshold would have rejected the null hypothesis, thus concluding that there is some evidence from our data set for an interaction effect.

Wald test (or z-test)

The likelihood ratio test is the recommended hypothesis test for comparing two nested logistic regression models. A different test (or **z-test**), called the Wald test, is often reported in scientific publications and is part of the default output of many popular software packages. For example, the information from the 'log odds' part of Table 2.11 (see above) may be displayed in a format as shown in Table 2.13.

Table 2.13 Coefficients and standard errors from Model 2.3, with Wald statistics and p-values (a common format of output tables from statistical analysis software)

	Coefficient	SE	Wald (z)	p (two-sided)
Intercept	−1.34	0.13	−10.47	0.000
Highschool	1.57	0.20	8.02	0.000
Youngkids	−0.85	0.36	−2.37	0.018
Highschool × Youngkids	0.82	0.47	1.77	0.078

The Wald tests displayed in each row are tests of the null hypothesis that the coefficient is zero in the population. This Wald statistic is calculated and evaluated like a z-statistic, that is

$$z = \frac{\hat{\beta}}{SE}$$

and a p-value is calculated by referring this z-statistic to the standard normal distribution. The Wald test can also be generalised to test hypotheses about multiple coefficients (e.g. testing that the coefficients for *Youngkids* and the interaction *Highschool × Youngkids* are both zero), but tests of multiple coefficients are not usually part of default software output, and I don't discuss them here. Such tests of multiple coefficients can also be conducted using the LRT, and in general the latter is the preferable method.

In most situations, the Wald test and the LRT produce very similar p-values and suggest the same conclusion. For example, in our case the Wald test for the interaction *Highschool × Youngkids* gives $p = 0.078$ (see Table 2.13), whereas the LRT gave $p = 0.070$. However, it has been shown that the LRT is more accurate than the Wald test, particularly in small samples or when the proportion of events is close to either 0 or 1. Thus, the LRT should be the preferred test,[xi] and in general I recommend not to rely on default software output, but to carry out LRTs instead.

Another disadvantage of displaying Wald tests of single coefficients by default is that it encourages the problematic habit of conducting hypothesis tests routinely and without thought, and of basing important modelling decisions (e.g. decisions about which variables to include in a model) purely or mainly on results of hypothesis tests, without considering the many ways in which this can result in misleading decisions. (I will say more about this in Chapter 6.)

Logistic regression: an example with multiple predictors

To illustrate how researchers might use logistic regression to investigate social science research questions, let's look at an example. Kraaykamp et al. (2010) analysed data from the Family Survey Dutch Population (FSDP) to investigate how cultural consumption varies with social status. In particular, they hypothesised that those with

---

[xi]This is exactly the same reason why I recommend profile likelihood confidence intervals over Wald confidence intervals (see the earlier section 'Confidence Intervals').

higher social status would be more likely to pursue highbrow cultural activities than those with lower social status. The sociological theory underlying this idea is that people of high social status seek to display this status by attending social events that in themselves have high prestige.[xii] One indicator of highbrow cultural activity that Kraaykamp et al. had available was classical concert attendance. The authors hypothesised that those with higher social status would be more likely than those with lower social status to attend classical concerts. For the sake of illustrating logistic regression, I will treat classical concert attendance as a binary variable, as I have done previously in this chapter.[xiii]

The main predictor of interest in this investigation is social status. In the FSDP data set, occupational status was measured via the International Socio-Economic Index (ISEI, Ganzeboom et al., 1992). The ISEI assigns a number to a person's occupation reflecting that occupation's social status, or prestige. A higher number indicates a higher status. The distribution of ISEI in the FSDP data set is illustrated in Figure 2.7. For the purpose of the analysis that follows, the ISEI score was **z-standardised**.

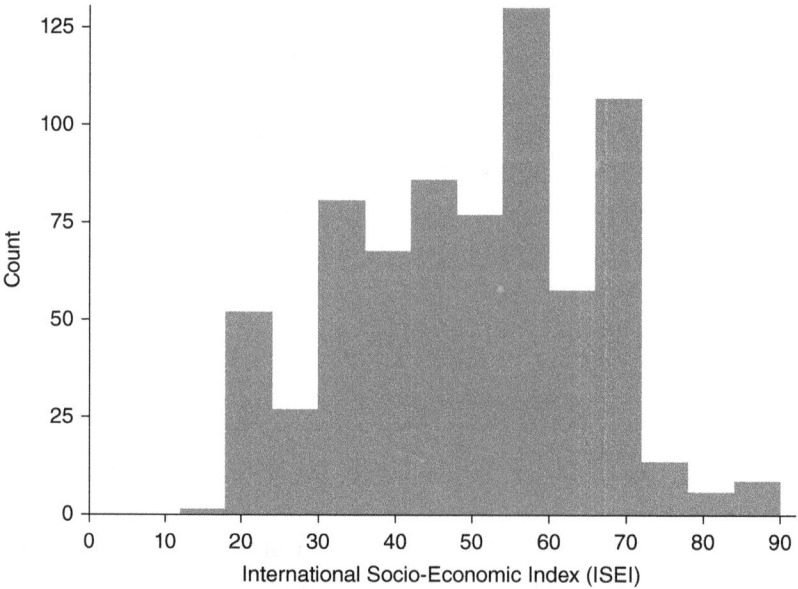

Figure 2.7 Histogram of the ISEI score in the Family Survey Dutch Population (n = 717)

[xii]An extreme example of this would be a person who dislikes classical music but nevertheless attends classical concerts in order to display their identity as a consumer of highbrow culture and to mingle with high status individuals.

[xiii]Kraaykamp et al. (2010) analysed the data differently and did not use binary logistic regression, but nonetheless, their research question and their data are useful for illustrating logistic regression.

ISEI is a continuous variable. For the moment, we will assume that the relationship between ISEI and the log odds of classical concert attendance is linear. Further below, we will investigate non-linearity between a continuous predictor and the log odds of the outcome event.

Now, to investigate the hypothesis that people of higher social status attend classical concerts to display their status, it may not be enough to simply estimate the association between social status and classical concert attendance. We may, in addition, wish to consider potential **confounding** variables. For example, income is a potential **confounder**. Tickets for many classical concerts are expensive, and people with high-status occupations are likely to earn more money than people with lower status occupations. Therefore, if social status is related to classical concert attendance, part of the reason might be that high-status individuals are more likely to be able to afford the price for the tickets. Thus, we may wish to control for household income in our logistic regression model. Similar arguments can be made for other potential confounders, of course.

Let's suppose we have carefully thought about which variables we wish to control for and have selected income, age, education, the presence of young children in the household, and area of residence (rural or urban). Kraaykamp et al. (2010) include many more predictors in their analysis, but we will keep the model small and simple for the purpose of this example. Our logistic regression model is as follows:

Model 2.5:

$$\ln\left(\frac{p_i}{1 - p_i}\right) = \beta_0 + \beta_1 Highschool_i + \beta_2 Youngkids_i + \beta_3 Rural_i +$$
$$\beta_4 AgeC_i + \beta_5 \log(income)_i + \beta_6 z.ISEI_i$$

where *Highschool*, *Youngkids* and *AgeC*[xiv] are defined as previously in this chapter, and

- *Rural* is a dummy variable identifying the type of the respondent's area of residence, coded 0 = urban and 1 = rural;
- Income is measured in Dutch guilders, the currency in the Netherlands at the time that the data were collected (1998). Data exploration, which I don't show here, indicated that the prediction of classical concert attendance could be improved by using log(*income*) in the model, rather than income. The log **transformation** is a common transformation for amount variables, such as variables that measure amounts of money (see also *The SAGE Quantitative Research Kit*, Volume 7, Chapter 3, for a detailed explanation and examples of log transformations). After taking the logarithm,

---

[xiv]Recall that *AgeC* is age centred at 40 years (see the section 'Logistic Regression With a Numeric Predictor').

I centred the variable around the (log of the) approximate median income (5000 guilders per month), so that the value 0 indicates a person with median income.

- *z.ISEI* is the International Socio-Economic Index, an indicator of occupational social status. For this analysis, I have z-standardised ISEI, such that its mean is 0 and its standard deviation is 1.

I have estimated Model 2.5 on the data from the FSDP. Table 2.14 shows the estimated coefficients, standard errors, odds ratios and 95% CIs for the odds ratios.

Table 2.14 Estimates from a logistic regression predicting classical concert attendance (Model 2.5)

	Coefficient	SE	OR	95% CI for OR
Intercept	−1.483	0.146		
Highschool	1.350	0.205	3.86	[2.59, 5.79]
Youngkids	−0.078	0.240	0.92	[0.57, 1.48]
Rural	−0.116	0.280	0.89	[0.51, 1.53]
Age (centred at 40)	0.034	0.009	1.03	[1.02, 1.05]
log(income) (centred)	0.825	0.256	2.28	[1.39, 3.80]
ISEI (standardised)	0.187	0.112	1.21	[0.97, 1.50]

Note. log(*income*) = log total household income in Dutch guilders per month, centred at median [log(5000 guilders)]. Data were taken from the Family Survey Dutch Population. CI = confidence interval; *OR* = odds ratio; *SE* = standard error.

The main research question concerns the effect of social status (ISEI) on classical concert attendance. The estimated odds ratio for *z.ISEI* is 1.21. This suggests that a 1-*SD* difference in ISEI is associated with about 21% higher odds of attending classical concerts, controlling for the other variables in the model. This estimate is in the direction that we expected given the research hypothesis (higher status makes classical concert attendance more likely). However, the 95% CI for the odds ratio is [0.97, 1.50]. That is, the data are consistent with the odds ratio being equal to 1 or even slightly smaller. An LRT of Model 2.5 against a model without ISEI gives $\Lambda = 2.80$, $df = 1$, $p = 0.094$. Overall, the findings suggest that we have some evidence from these data for an effect of social status on classical concert attendance, controlling for the other variables in the model, but that this evidence is not very strong.

We will look at other ways to interpret and visualise the results from a logistic regression with many predictors in the section 'Interpretation of Effect Sizes and Graphical Illustration'. Before we do so, we should evaluate the quality of our model and investigate whether it meets its statistical assumptions. That is the topic of the next section.

Model evaluation

This section is devoted to assessing the *goodness of fit*, investigating assumptions and evaluating the overall quality of a logistic regression model. The techniques for doing so are different from the techniques we use for the same purposes in the context of linear regression (see *The SAGE Quantitative Research Kit*, Volume 7). In linear regression, we use residual diagnostics to assess model fit and the plausibility of model assumptions, and we can calculate the coefficient of determination (R^2) as a summary measure of how well the model predicts the outcome. In logistic regression, residual diagnostics tend to be less useful, as the next section ('Residuals in Logistic Regression') explains. Instead, we use graphical techniques and statistical tests to assess the quality of what is called the *calibration* of a logistic regression model. These techniques are discussed in the sections on 'Model Calibration: Graphical Exploration' and 'The Hosmer–Lemeshow Test of Model Calibration'. Indices of prediction quality for logistic regression models are considered in the section 'Model Quality Indices for Logistic Regression'.

Residuals in logistic regression

In general, in a logistic regression with continuous predictors, most or all individuals will have different predicted probabilities in the range between 0 and 1, but each individual can have only one of two possible outcome events: either $Y = 0$ or $Y = 1$. So by definition an individual's actual outcome differs from the prediction.

The (raw, unstandardised) residual from a logistic regression model is

$$e_i = Y_i - \hat{p}_i$$

where e_i is the raw residual of the *i*th person; Y_i is the outcome of the *i*th person, coded as either 1 or 0; and $\hat{p}_i$ is the model's predicted probability of $Y_i = 1$ for the *i*th person.

Because Y_i can only be either 0 or 1, the residual is always either $1 - \hat{p}_i$ or $0 - \hat{p}_i$. This leads to a peculiar pattern of the residuals, which is illustrated in the top panel of Figure 2.8: when plotted against the predicted probabilities, the residuals form two parallel lines, with the upper line consisting of people with the event $Y_i = 1$ (here, people who attend classical concerts) and the lower line consisting of people without the event ($Y_i = 0$). Note also that the raw residual must lie between −1 and +1. Outlying residuals, in the sense that we encountered in linear regression, therefore do not occur in logistic regression.

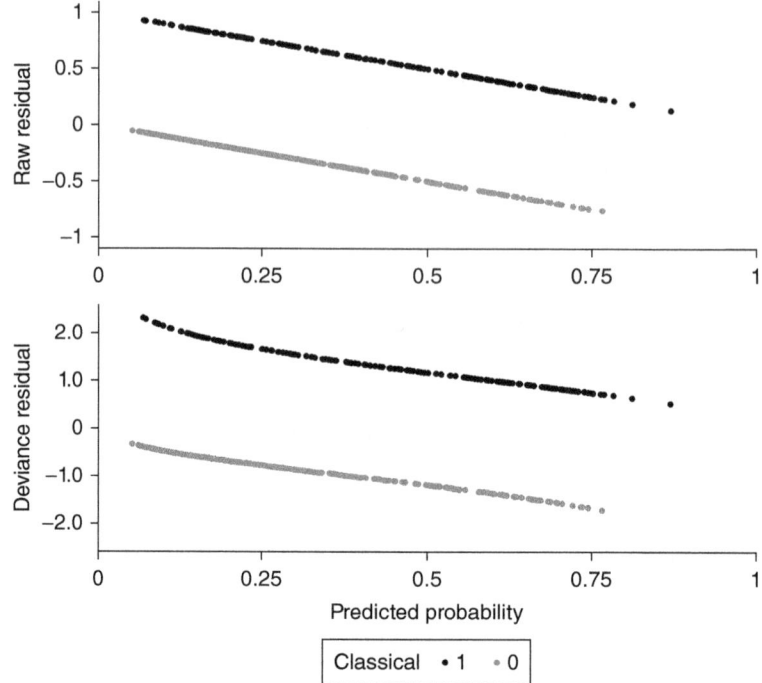

Figure 2.8 Raw and deviance residuals by predicted probabilities from Model 2.5

Other types of residuals have been developed. For example, the bottom panel of Figure 2.8 shows the *deviance residuals*, which are derived from the raw residuals by scaling them so that, in a well-fitting model, they have approximately a standard normal distribution. You can find more detail on different types of residuals for logistic regression, and how they have been used, in Hilbe (2009) and Agresti (2013).

On the whole, the inspection of residuals is less useful in logistic regression compared to linear regression. Instead of investigating residuals of individual observations, it is often more useful to consider groups of observations, as we will see in the next section.

Model calibration: graphical exploration

Before we use a model to draw substantive conclusions or make predictions, we need to verify that it is well calibrated, such that the relationship between continuous predictors and the probability of the outcome follows the shape implied by the model. This is akin to checking the assumption of linearity in linear regression. In a well-calibrated model, the model accurately predicts the probability of the outcome event for everyone in the population for which it was developed. In a poorly calibrated model, the prediction may fit some groups in the population better than others.

One way to inspect model calibration is to divide our sample into equal-sized groups according to the predicted probabilities. For example, we might divide the sample into decile groups. Then the first group contains the 10% of cases with the lowest predicted probabilities, the second group contains the 10% of cases with the next lowest probabilities, and so forth.

For each of these 10 groups, we then calculate the average predicted probability of having the outcome event and compare it with the observed proportion of sample members who actually had the event. This comparison is displayed in Table 2.15, using the predictions from Model 2.5. If the model is well calibrated, the mean predicted probability of classical concert attendance should be similar to the observed proportion of classical concert goers in each group.

Table 2.15 Comparing the observed and expected numbers of respondents who attend and don't attend classical concerts, by decile group of predicted probabilities (from Model 2.5)

Decile Group	N	Mean Predicted Probability	Expected Number		Observed Number		
			Attends Classical Concerts	Does Not Attend	Attends Classical Concerts	Does Not Attend	Observed Proportion
1	72	0.085	6.1	65.9	8	64	0.111
2	72	0.116	8.4	63.6	6	66	0.083
3	71	0.145	10.3	60.7	10	61	0.141
4	72	0.175	12.6	59.4	12	60	0.167
5	72	0.214	15.4	56.6	19	53	0.264
6	71	0.265	18.8	52.2	14	57	0.197
7	72	0.358	25.8	46.2	28	44	0.389
8	71	0.471	33.4	37.6	35	36	0.493
9	72	0.586	42.2	29.8	41	31	0.569
10	72	0.707	50.9	21.1	51	21	0.708

Note. Data were taken from the Family Survey Dutch Population.

Figure 2.9 facilitates the same comparison by plotting the observed proportions of classical concert attenders in each of the 10 groups from Table 2.15 against the mean predicted probability in each group. The diagonal line indicates perfect agreement: if the model was perfectly calibrated, all dots would be on this line. Large deviations from the line may indicate poor model calibration, particularly if the deviations form a non-linear pattern. In Figure 2.9, there are some deviations, maybe indicating that our model has not perfectly captured the shape of the relationship between classical concert attendance and one or several predictors. However, there is no easily recognisable pattern to the deviations, so that I don't see how we could derive a better model based on these data alone. In social science data, plots like Figure 2.9 often look slightly messy and ambiguous. For clarity, Figure 2.10 shows two ideal typical situations, one of very good calibration and another of poor calibration with an obvious pattern of deviations.

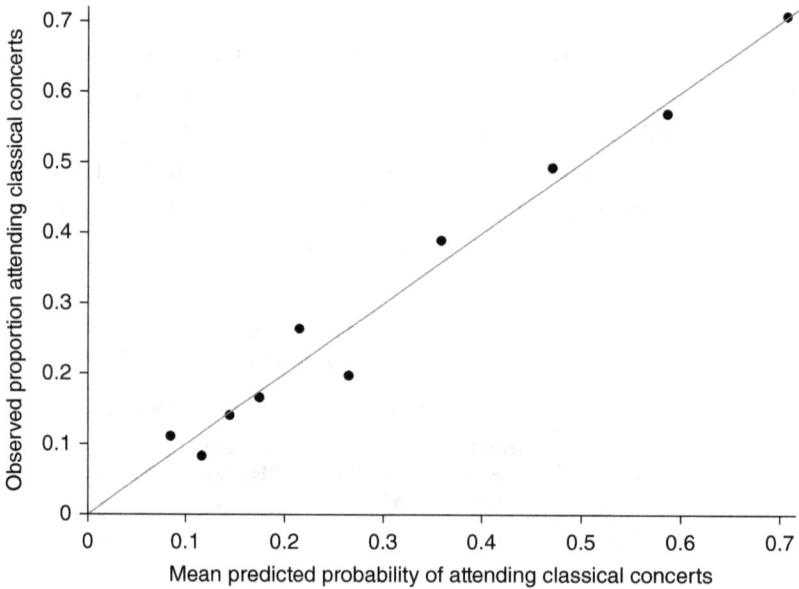

Figure 2.9 A graphical check of model calibration for a logistic regression: observed proportions of individuals who attend classical concerts by mean predicted probability (from Model 2.5), by decile group

The model at the top of Figure 2.10 is well calibrated: the observed proportions of individuals with the event match very closely the average predicted probabilities for decile groups. In contrast, the model at the bottom is poorly calibrated. Not only do the observed proportions and the predicted probabilities not match, there is also a pattern to the deviations. The dots approximately form a parabola, or U-shape. Since U-shaped relationships can often be modelled by employing a square transformation, in this situation we might investigate whether a square transformation of one of the predictor variables might help to improve the model fit (see also *The SAGE Quantitative Research Kit*, Volume 7, Chapter 3).

The Hosmer–Lemeshow test of model calibration

In addition to the graphical exploration, we can test model calibration formally. A popular test was first proposed by Hosmer and Lemeshow (cf. Hosmer et al., 2013). The Hosmer–Lemeshow test is a goodness-of-fit test that assesses the null hypothesis that our model is perfectly calibrated. The test works by dividing the sample into

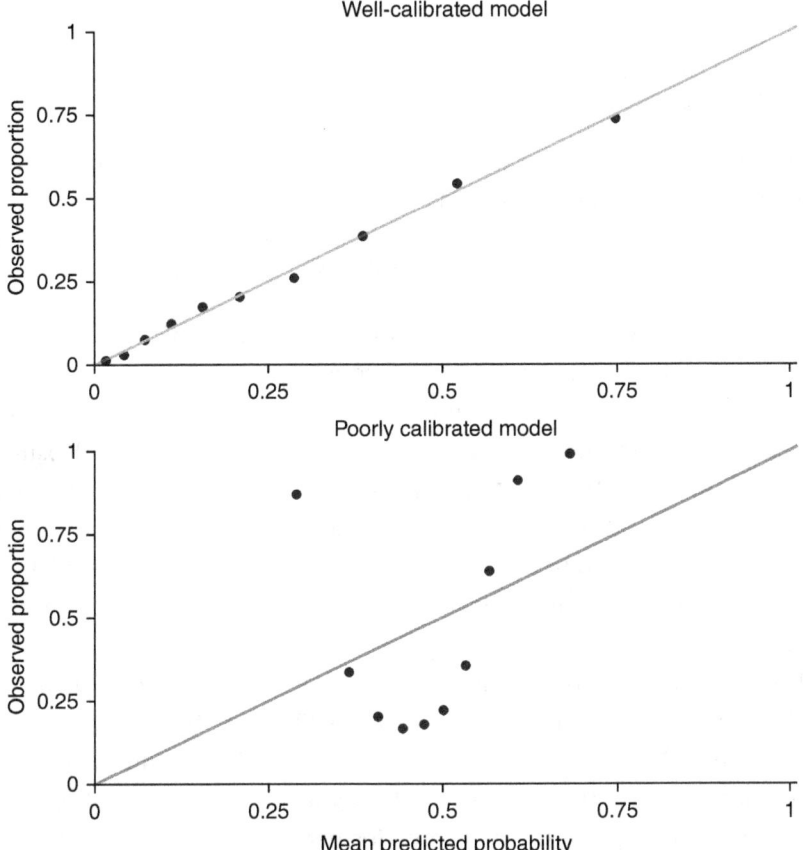

Figure 2.10 Hypothetical examples of a well-calibrated and a poorly calibrated logistic regression model

groups according to predicted probabilities, as we did in the previous section. Let's call the number of groups g. The Hosmer–Lemeshow test compares the observed with the expected numbers of events and non-events in the g groups. For example, in Table 2.15, we divided the sample into $g = 10$ groups and displayed the expected and observed numbers of people who attend and do not attend classical concerts in each group. The Hosmer–Lemeshow test statistic, which I will call H,[xv] is calculated from these numbers by the following formula:

$$H = \sum_{k=0}^{1} \sum_{m=1}^{g} \frac{(O_{kl} - E_{km})^2}{E_{km}}$$

[xv]Hosmer and Lemeshow called their statistic C. I call it H instead, to avoid confusion with the C-statistic that is discussed in the following section, 'Model Quality Indices for Logistic Regression'.

where

- k is equal to either 0 or 1, denoting either non-events (0) or events (1),
- $m = 1, 2, 3, \ldots, g$ denotes the group, where the number of groups is g (e.g. $g = 10$ in Table 2.15),
- O_{0m} denotes the number of observed non-events (outcome = 0) in the mth group,
- O_{1m} denotes the number of observed events (outcome = 1) in the mth group,
- E_{0m} denotes the number of expected non-events in the mth group, and
- E_{1m} denotes the number of expected events in the mth group.

If the observed numbers of events (and non-events) match the expected numbers exactly, then the Hosmer–Lemeshow statistic is zero. If there are big divergences between the observed and expected numbers of events (and non-events), then the Hosmer–Lemeshow test statistic is large. Hosmer and Lemeshow (cf. Hosmer et al., 2013) have shown that their statistic approximately follows a chi-squared distribution with $df = g - 2$, if the model is well calibrated, if $g \geq 6$, and if the proportion of events is between 0.1 and 0.9. If the p-value calculated from the test statistic is small (smaller than 0.05, say), then this constitutes evidence against the model being well calibrated.

The Hosmer–Lemeshow test statistic for our example Model 2.5, calculated from the numbers in Table 2.15, turns out to be $H = 6.43$. Using the chi-squared distribution with $df = 10 - 2 = 8$, this yields a p-value of $p = 0.599$. This indicates that there is no strong evidence, from this test, against the assumption of good calibration.

Like all statistical hypothesis tests, the Hosmer–Lemeshow test should be interpreted in the context of the data and should not be the only criterion by which we judge the calibration of the model. It is advisable to employ graphical diagnostics, as shown in Figure 2.9, in addition to the Hosmer–Lemeshow test. In particular, note the following:

1   If the sample size is small, the Hosmer–Lemeshow test may lack statistical power to detect poor calibration. In that case, if we blindly trust the p-value, we may erroneously conclude that we have a well-calibrated model, when in fact we do not.
2   On the other hand, when the sample size is very large, the test can be sensitive to small differences between observed and predicted values, and we may thus mistakenly conclude that we cannot trust the model when in fact it is 'good enough'. Paul et al. (2013) argue that the Hosmer–Lemeshow test should not be used for samples with $n > 25,000$.
3   The choice of the number of groups, g, is somewhat arbitrary, and the results of the test can differ depending on the choice of g. Moreover, the choice of g can also affect the power of the test. Paul et al. (2013) recommend choosing $g = 10$ for sample sizes up to 1000. For larger sample sizes, they recommend choosing g depending on the sample size as well as the number of events in the sample. They give a formula to determine what they consider the optimal g (Paul et al., 2013), but further methodological research may well lead to changes to these recommendations in the future.

Model quality indices for logistic regression

It can be advantageous to have a statistic that summarises the prediction quality of a logistic regression model. In linear regression, the R^2 statistic serves as such a measure, being an estimate of the proportion of outcome variance explained by the predictors. In logistic regression, no exact equivalent of the R^2 statistic is available. Statisticians have developed several measures of the quality of a logistic regression model, but there is no universally accepted standard. Different researchers use different measures, and statistical software packages differ in what they show as default output. You will also find that different scientific disciplines have different fashions regarding which statistics to report when summarising the quality of a logistic regression model. As you will see in this section, I recommend some of these statistics more than others.

Discrimination: the C-statistic: The C-statistic measures the discrimination of a logistic regression model. It measures how well the model discriminates between cases with the outcome event and those without (e.g. between the ill and the healthy, or between those who attend classical concerts and those who do not). For sensible models, the C-statistic varies between 0.5 and 1. A higher C-statistic indicates better discrimination. $C = 0.5$ indicates a useless model, while $C = 1$ indicates a model that discriminates perfectly between those with the event and those without.[xvi] Model 2.5 has a C-statistic of 0.761, indicating moderate discrimination.

The C-statistic can be calculated as follows. Consider a pair of people, person A and person B, from our data set. The pair is chosen such that person A has the characteristic we are trying to predict (they are ill, a concert goer, etc.), while person B does not. Our pair is called concordant if person A has a higher predicted probability of having the characteristic than person B. Now, we can list all pairs of cases from our data set where one member of the pair has the event, while the other does not. The C-statistic is the proportion of concordant pairs among this list. Thus, the C-statistic can be considered an estimate of the probability that any random pair of cases chosen as above is concordant.

Pseudo-R^2 measures: Recall (from Volume 7 of *The SAGE Quantitative Research Kit*) that the coefficient of determination, R^2, is the proportion of outcome variance accounted for by a linear regression model. R^2 can also be interpreted as the square of the correlation between the observed outcome values and the predicted values. For a logistic regression model, a straightforward equivalent of the R^2 statistic does not exist, and different approaches to calculating R^2, which in a linear regression would

---

[xvi]When $C = 1$, we can find a cut-off point for the predicted probability such that everyone whose predicted probability is above the cut-off has the event, and everyone whose predicted probability is below the cut-off does not have the event.

all yield the same result, would lead to different results if applied to logistic regression. However, several different R^2-like measures have been developed for logistic regression, with the aim of having a single number indicating how well our model predicts the binary outcome. These measures are referred to as *pseudo-R^2 statistics*. There are several of these. They often give very similar values, but it is unclear whether some are superior to others. To give just one example, McFadden's R^2 statistic is defined as follows:

$$R^2_{\text{McFadden}} = 1 - \frac{\text{LL}_{\text{Model}}}{\text{LL}_{\text{Null}}}$$

where LL_{Model} is the log likelihood of the model we are evaluating, while LL_{Null} is the log likelihood of a null model without any predictors. (The null model must be estimated using the same data set and the same outcome, of course.) McFadden's R^2 can theoretically take values between 0 and 1, although it can never exactly equal 1. A value of 0 indicates that the predictors in our model make no contribution to the prediction of the outcome, while a value close to 1 indicates very good prediction of the outcome.

McFadden's R^2 can be used to compare nested and non-nested models on the same data set. To compare models with different numbers of parameters, it might be better to use McFadden's adjusted R^2:

$$R^2_{\text{McFadden adj.}} = 1 - \frac{\text{LL}_{\text{Model}} - K}{\text{LL}_{\text{Null}}}$$

where K is the number of parameters in the model; in other words, K is the number of coefficients (intercept and slopes) that are being estimated.

Several other pseudo-R^2 measures have been developed (see Long & Freese, 2014, for an overview). In general, the pseudo-R^2 statistics do not have a neat interpretation. They cannot be interpreted as the proportion of variance explained, for example, as the R^2 for linear regression can be. Thus, many statisticians consider the pseudo-R^2 measures of little value. For our Model 2.5, $R^2_{\text{McFadden}} = 0.157$, and $R^2_{\text{McFadden adj.}} = 0.141$. Taken by themselves, it is not clear whether these values represent a 'good model', a 'poor model' or something in between.[xvii]

[xvii]For comparison of nested models, likelihood ratio tests are likely to be more informative than (adjusted) pseudo-R^2 statistics. For the comparison of non-nested models, an adjusted pseudo-R^2 might be used, although it is probably more advisable, and more common, to use information criteria instead. Two information criteria, the AIC (Akaike information criteria) and the BIC (Bayesian information criteria), are introduced in Chapter 5 of this book.

Interpretation of effect sizes and graphical illustration

Once we are happy that our model is well calibrated and discriminates reasonably well between those with the outcome event and those without, we might wish to think about how to interpret the results. Let's consider again Model 2.5, and the results displayed in Table 2.14.

Recall that Kraaykamp et al. (2010) used the FSDP data to investigate the hypothesis that the higher the social status, the higher the probability of attending classical concerts, controlling for other variables. If the data bear out this hypothesis, this would be consistent with the theory of social distinction through cultural consumption. Kraaykamp et al. (2010) used a larger data set and more predictors[xviii] than I have included in Model 2.5., but we will consider the results from my simple model to illustrate the interpretation of coefficients.

From Table 2.14, we see that the estimated odds ratio for 'ISEI', the z-standardised social status variable, is 1.21, with a 95% CI from 0.97 to 1.50. So in this data set, a 1-*SD* difference in social status is associated with 21% increased odds of classical concert attendance, controlling for all other variables in the model. Since the 95% CI includes odds ratios of 1 and slightly below, we should be cautious in drawing firm conclusions from our result, since it is reasonably plausible that the data may have come from a population where there is no relationship (or a very small relationship) between social status and classical concert attendance, when controlling for the other variables.

In general, when our research question focuses on the associations between one or several predictors and the outcome, it is often advantageous to illustrate the modelled relationships. This can help us understand the strengths of the relationships of interest. For example, let's assume we wish to understand in more detail how classical concert attendance varies with social status according to Model 2.5, and how the strength of this relationship compares with the effects of other predictors.[xix] Then we may construct a graph such as Figure 2.11.

[xviii]Kraaykamp et al. (2010) used data from several surveys conducted in the Netherlands between 1992 and 2003 and also used information from both the primary survey respondents and their married or cohabiting partners. I present here a simplified analysis using the 1998 survey and primary respondents only to make the example easier to explain.

[xix]There is nothing wrong with doing this, even though we have found that there is only modest statistical evidence ($p = 0.094$) from these data to support the idea that social status is related to classical concert attendance. However, if we do include such a graphical illustration in our research report, we of course also should report the confidence interval for the relevant odds ratio, and possibly the *p*-value for a test of the null hypothesis of 'no effect', to make clear the strength (or the weakness) of the evidence for the effect. I should also add that Kraaykamp et al. (2010), who use a larger data set, a larger set of covariates and a different regression model, do find evidence of a relationship between social status and classical concert attendance.

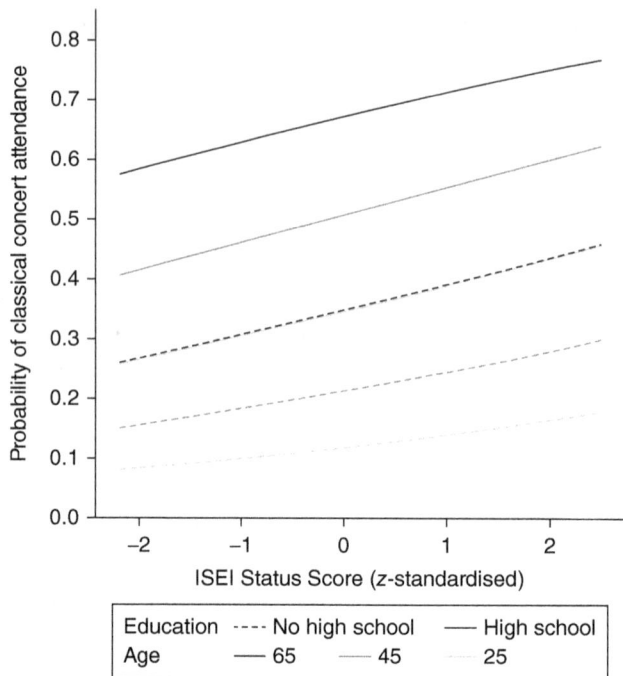

Figure 2.11 Predicted probabilities of classical concert attendance from Model 2.5

Note. Predicted probabilities are given for people without young children in the household (*Youngkids* = 0), living in an urban area (*Rural* = 0), and with median income [centred log(*income*) = 0]. Data were taken from the Family Survey Dutch Population. ISEI = International Socio-Economic Index.

Figure 2.11 shows how the predicted probability of classical concert attendance varies with age, education and ISEI score. In order to calculate these predicted probabilities, I have held the three other variables in Model 2.5 constant, by setting *Youngkids* = 0, *Rural* = 0, and centred log(*income*) = 0. This means that the predicted probabilities displayed in Figure 2.11 apply to those who don't have young children in the household, live in an urban area, and have a monthly household income of 5000 Dutch guilders. (Recall that the log-transformed income variable was centred on this value, which is approximately the median income in this sample.) I could have chosen to set these variables to different values. Then the predicted probabilities would be different, but the shapes of the relationships would be very similar to those that we see in Figure 2.11.

Figure 2.11 shows the predicted probabilities for the whole range of (*z*-standardised) ISEI, by educational group and for three example ages, representing the young, the middle-aged and older people, respectively. (I could have chosen other example ages, of course.)

The plot illustrates that having a high school education, being older and having a higher status occupation are all associated with a higher probability of attending classical concerts in this data set. The plot also allows us to assess the strengths of the relationships estimated by our model. For example, we see that the difference between those with high school education and those without is larger than the difference between those at the top of the ISEI status scale and those at the bottom. Note also the two lines that are almost overlapping: 25-year-olds with high school education have about the same predicted probability as 65-year-olds without high school education. That is, according to our model, 40 years of age difference are associated with about the same difference in the probability of attending classical concerts as a high school education.

Things that might go wrong: estimation problems

There are some situations in which the estimation of a logistic regression model may be unreliable or even impossible. This can occur if the number of events or non-events is small relative to the number of predictors that we wish to include in the model. In particular, we may encounter estimation problems when

- our sample is very small,
- the probability of the outcome event is very low (or very high),
- we have a large number of predictors in our model, or
- two or all three of the above.

Consider, for example, a disease that occurs in 1% of the population – that's a low outcome event probability. Let's assume we wish to study predictors of this disease, and that we take a random sample of 1000 people to do so. Then the expected number of people with the disease in our sample is 10 (1% of 1000).

In such a situation, we may encounter a problem called *sparseness*: this means that for a particular combination of predictor values, there are few or no events. To continue the example above, there may be very few young people in our sample who have the disease. This may cause the estimate of the coefficient of age to be very uncertain. Such uncertainty is indicated by large standard errors around slope coefficients and leads to very wide confidence intervals.

A more extreme case of sparseness occurs when a predictor category leads to *perfect prediction* in the sample data set. For example, no men in the sample may have the disease (although in the population, some men do have it). Perfect prediction leads to estimation problems, and statistical software may return an error or a warning, or calculate the standard errors to be of gigantic size.

To guard against fitting models based on too sparse data, researchers have suggested considering the number of *events per variable* (EPV): this is the ratio of the number of events (people with $Y = 1$, assuming that $Y = 1$ is rarer in the sample than $Y = 0$) over the number of predictors. As a rule of thumb, it has been suggested that the EPV should be at least 10 for logistic regression to be appropriate. However, methodological research has found that this rule of thumb is not sufficient, and that accurate model estimation cannot necessarily be guaranteed even with EPV = 20 (Courvoisier et al., 2011; Van Smeden et al., 2016).

To illustrate the issues, consider Table 2.16, which shows data from the British Cohort Study 1970, a study of people born in a week in March 1970.[xx] In the year 2000, when the cohort members were 30 years old, they were asked whether they had ever suffered from bulimia or from a problem with compulsive eating. Bulimia is an eating disorder characterised by bouts of binge eating and subsequent efforts to purge the body, such as by vomiting or use of laxatives. Identifying predictors of bulimia might help scientists to better understand its aetiology. The example takes inspiration from a piece of research by Nicholls et al. (2016).

Table 2.16 Bulimia (age 30) by gender and mother's education in the British Cohort Study 1970

Gender	Bulimia	Mother's Education				Total	
		Below A Level		A Level or Higher		Total	
Female	No ED	3865	(98.56%)	522	(97.57%)	4287	(98.44%)
	Bulimia	55	(1.44%)	13	(2.43%)	68	(1.56%)
	Total	**3820**	**(100.00%)**	**535**	**(100.00%)**	**4355**	**(100.00%)**
Male	No ED	3638	(99.84%)	511	(99.80%)	4149	(99.83%)
	Bulimia	6	(0.16%)	1	(0.20%)	7	(0.17%)
	Total	**3644**	**(100.00%)**	**512**	**(100.00%)**	**4156**	**(100.00%)**
Total	No ED	7403	(99.18%)	1033	(98.66%)	8436	(99.12%)
	Bulimia	61	(0.82%)	14	(1.34%)	75	(0.88%)
	Total	**3644**	**(100.00%)**	**512**	**(100.00%)**	**8511**	**(100.00%)**

Note. Respondents who reported anorexia are excluded from this table. 'Bulimia' encompasses respondents reporting bulimia or compulsive eating. Data were taken from the British Cohort Study 1970. ED = eating disorder.

The total sample size in Table 2.16 is 8511 (4355 females and 4156 males). This is not a small sample by most standards. The number of sample members reporting bulimia is 75 (68 females and 7 males), representing 0.88% of the total sample. Thus, bulimia is a relatively rare diagnosis for people in this cohort (at least as far as they report it). Although we have a reasonable number of females with bulimia, we have only 7 males. Thus, it will be difficult to estimate the influence of further predictors on the probability of bulimia among males specifically.

[xx]See: https://cls.ucl.ac.uk/cls-studies/1970-british-cohort-study/

To illustrate, consider a second categorical predictor, mother's education. It turns out that only one man whose mother had an A-level education or higher reports bulimia. If we assume that the effect of mother's education on the risk of bulimia is the same for females and males, this is not a big problem, since overall we have 14 bulimia cases among children (male and female) of highly educated mothers. However, if we suspect that mother's education might have a different relationship with the risk of bulimia among males than among females, this data set is not large enough to investigate the issue.

For example, suppose that we wished to estimate the model:

Model 2.6:

$$\ln\left(\frac{p_i}{1 - p_i}\right) = \beta_0 + \beta_1 Female_i + \beta_2 MumAlev_i + \beta_3 Female_i \times MumAlev_i$$

Fitting this model to the data summarised in Table 2.16, we obtain the estimates shown in Table 2.17.

Table 2.17 Coefficient estimates, standard errors and confidence intervals for the prediction of bulimia by gender, mother's education and their interaction (Model 2.6)

	Coefficient	SE	OR	95% CI for OR
Intercept	−6.41	0.41		
Female	2.18	0.43	8.86	[4.13, 23.03]
MumAlev	0.17	1.08	1.19	[0.06, 6.97]
Female × MumAlev	0.36	1.13	1.44	[0.22, 28.42]

Note. 95% confidence intervals were computed using the profile likelihood method. OR = odds ratio; CI = confidence interval; SE = standard error.

The signals that should alert us to the fact that there is a problem here are the large standard errors, particularly for *MumAlev* and the interaction *Female × MumAlev*. These large standard errors in turn result in very wide confidence intervals. For example, the confidence interval for *MumAlev*, which measures the effect of mother's education on the bulimia risk for males, is [0.06, 6.97]. Thus, the model estimates that mother's education is associated with somewhere in between a 94% reduction in odds up to a seven-fold increase in odds. This is a grotesquely wide confidence interval, indicating that essentially we cannot say anything about the effect of mother's education on bulimia risk among men from this data set and this model.

In general, large standard errors and wide confidence intervals may be a sign that you have a sparseness problem in your data, and that you may not be able to estimate all the coefficients in your model. In this example, the data set is not sufficiently large to estimate the interaction of mother's education with gender.

In general, if you do encounter sparseness, what can be done?

- *Change the outcome:* Sometimes it might be possible to change the definition of the outcome such that the outcome event is less rare. For example, instead of trying to predict bulimia specifically, we may instead decide to predict 'any mental illness'. Of course, this changes the aims of the investigation.
- *Concentrate on key predictors or key subgroups:* If a particular predictor causes the sparseness problem, consider if it can be left out. For example, if very few males in your sample are recorded to have an eating disorder, but you are concerned that other predictors may not have the same relationship with the outcome for females and males, then consider conducting the investigation on females only. This changes the population to which we can generalise results.
- *Consider evidence from other data sets:* If the data are too sparse to model a certain predictor or term of interest, such as an interaction, consider evidence from other data sets to inform you whether the term is likely to be necessary or not. For example, if I was confident that the predictors of bulimia were the same among men and women, I might decide that I can use data from both men and women, but without fitting any interactions of gender with other predictors.

If none of the above actions are possible, or if they don't alleviate the problem, you may wish to consider a descriptive analysis without a statistical model. Sometimes it may even be better not to do a statistical analysis, if the data set is not sufficiently large, or not of sufficient quality, to estimate a meaningful model in whose estimates we can have confidence. In that case, it might be preferable to instead plan a new study that collects a larger and higher-quality data set.

Logistic regression in action

In this section, we shall walk through a complete logistic regression analysis, using many of the techniques introduced in the chapter so far. The purpose is to further illustrate how logistic regression might be used in practice by researchers to address a social scientific research question. We return to the Family Survey Dutch Population (FSDP) and the topic of cultural consumption. In this survey, the respondents were asked not only about their attendance of classical concerts but also about their attendance of pop music concerts. The theory of social distinction posits that classical concert attendance might partly have the social function of displaying an individual's social status. In this regard, pop concerts provide an interesting comparison, because they are not considered a highbrow cultural activity and are therefore unlikely to be used to display high social status. Thus, social distinction theory would hypothesise that the predictors of classical and pop concert attendance would be different, and in particular that high social status should not be positively associated with attendance at pop concerts. Thus, it might be informative to find out whether the predictors of attending classical and pop concerts are different or the same. To do so, we will consider the same set of predictors with which we modelled classical concert attendance in Model 2.5.

Before fitting a regression of pop concert attendance on our predictors, we should investigate our data in order to detect potential problems with sparseness, and to explore the shape of the relationship between continuous variables and the probability of pop concert attendance. In our case, 321 out of 799 respondents, or 40%, report attending pop concerts. Inspection of cross-tabulations of pop concert attendance with our categorical predictors (not shown here) suggests that there is no problem with sparseness. The relationship between social status (measured by the z-standardised version of the ISEI score, as in Model 2.5) and the log odds of pop concert attendance appears to be approximately linear, and the same goes for the relationship between log(*income*) and the log odds of pop concert attendance (explorations not shown). However, the case is more complicated for age, as we will see next.

Exploring the relationship between age and attending pop concerts

Even though the relationship between age and pop concert attendance may not be our primary research question, we wish to make sure that our model adequately

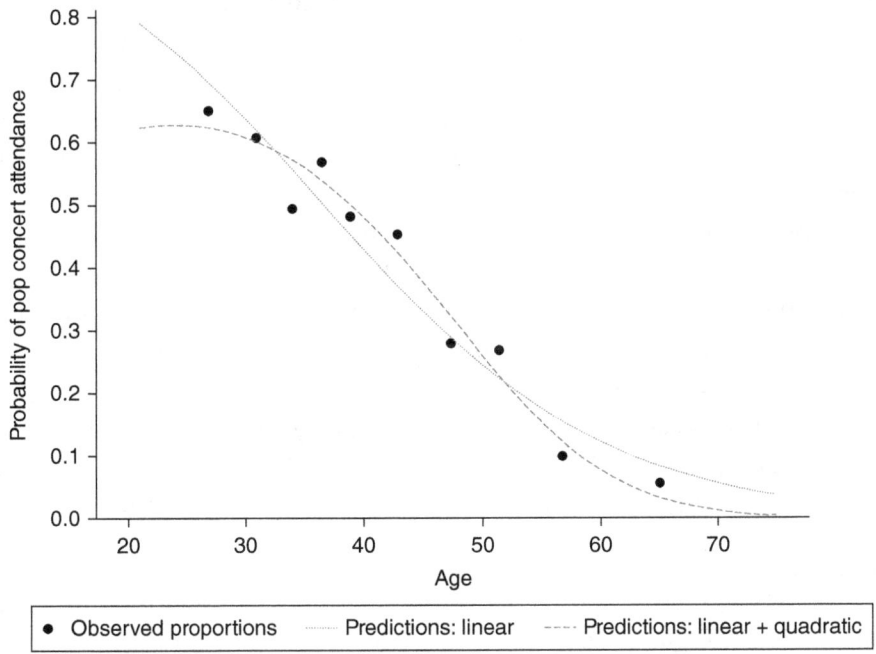

Figure 2.12 Observed proportions and predicted probabilities of pop concert attendance predicted by age: linear (Model 2.7) and linear + quadratic (Model 2.8)

represents the shape of this relationship. If the relationship between the outcome and a predictor is misspecified, this not only causes the coefficient estimate of this predictor to be misleading, but may also cause estimates of slope coefficients relating to other variables to be biased. Figure 2.12 shows the relationship between age and the probability of pop concert attendance. I have split the sample into 10 groups according to their age and plotted the mean age in each group against the observed proportion of pop concert attendance. It is clear that there is a strong negative relationship between age and pop concert attendance: the older the respondent, the less likely they are to attend pop concerts. The observed proportions seem to be roughly consistent with the inverted S-shape assumed by logistic regression for a negative relationship, but it's difficult to be sure about this from the picture.

Figure 2.12 also shows the predictions from two models. These are two potentially plausible models I chose to explore for the relationship between age and pop concert attendance. The first model is a simple logistic regression of pop concert attendance on age. This uses only a linear term for age (where linear means: linear in the log odds, i.e. the age variable is not transformed). This simple model is

Model 2.7:

$$\text{logit}(Pop) = \alpha + \beta_1 AgeC_i$$

where $AgeC$ is the mean centred age variable, calculated as $AgeC$ = age – 40 years.

I won't show the estimated coefficients from this model here, but the predicted probabilities derived from the estimation of Model 2.7 on the FSDP data are shown in Figure 2.12 as a dotted line. The predictions seem to fit the observed values reasonably well, but it looks as though this model might overestimate pop concert attendance for those under 35 and those over 55, while tending to underestimate it for the middle-aged. Putting this in more general terms, it seems that there is a systematic error in the predictions: overestimation in the tails of the predictor, and underestimation for values near the mean of the predictor. This suggests some unmodelled non-linearity in the relationship between the predictor and the log odds of the outcome event. This unmodelled linearity roughly follows a U-shape (high at each end, low in the middle).

If we suspect a U-shaped non-linearity, a candidate method for improving the model fit is the square transformation, $t(AgeC) = AgeC^2$. We calculate the square of mean-centred age and add it to our data set. We then estimate the model:

Model 2.8:

$$\text{logit}(Pop) = \alpha + \beta_1 AgeC_i + \beta_2 AgeC_i^2$$

That is, we posit that the probability of pop concert attendance depends both on (mean-centred) age and its square. The predictions from Model 2.8 are shown in

Figure 2.12 as a dashed line. This suggests that Model 2.8 may fit the observed proportions a little better than Model 2.7. The differences are subtle, but the predictions from Model 2.8 tend to be a little closer to the observations than the predictions from Model 2.7. Also, the deviations of Model 2.8's predictions from the observations do not appear to have a systematic pattern, such as the U-shaped pattern we discerned in the deviations associated with the predictions from Model 2.7.

To confirm our impression, we may conduct a likelihood ratio test of the two models. The null hypothesis is that Model 2.7 is sufficient to describe the relationship between age and pop concert attendance. The alternative hypothesis is that Model 2.8, which adds the square transformation, improves the prediction. The LRT gives $\Lambda = 8.095$, $df = 1$, $p = 0.0044$. This is strong statistical evidence, from these data, that the linear + quadratic Model 2.8 results in a better prediction of pop concert attendance than Model 2.7. Thus, we have a reason to include both a linear and a quadratic term for age in our full model predicting pop concert attendance.

Logistic regression of pop concert attendance on six predictors

Our model of interest is

Model 2.9:

$$\text{logit}(Pop) = \beta_0 + \beta_1 Highschool_i + \beta_2 Youngkids_i + \beta_3 Rural_i +$$
$$\beta_4 AgeC_i + \beta_5 AgeC_i^2 + \beta_6 \log(income)_i + \beta_7 ISEI_i$$

The results of estimating Model 2.9 on the FSDP data are displayed in Table 2.18.

Table 2.18 Estimates from a logistic regression predicting pop concert attendance (Model 2.9)

	Coefficient	SE	OR	95% CI for OR
Intercept	0.228	0.147		
Highschool	−0.262	0.204	0.77	[0.52, 1.15]
Youngkids	−0.617	0.203	0.54	[0.36, 0.80]
Rural	−0.432	0.256	0.65	[0.39, 1.07]
Age (centred at 40)	−0.087	0.010	0.92	[0.90, 0.93]
AgeC²	−0.0021	0.0009	0.9979	[0.9961, 0.9995]
log(income) (centred)	0.453	0.247	1.57	[0.97, 2.56]
ISEI (standardised)	0.104	0.103	1.11	[0.91, 1.36]

Note. 95% confidence intervals were computed using the profile likelihood method. log(*income*) = log total household income in Dutch guilders per month, centred at median [log(5000 guilders)]. I have given coefficients for *AgeC²* to four decimal points because otherwise rounding error for these coefficients would have been unduly large. Data were from the Family Survey Dutch Population. *SE* = standard error; *OR* = odds ratio; *CI* = confidence interval; ISEI = International Socio-Economic Index.

Before we assess what answer to our research question these results imply, we assess whether the model is well calibrated. The Hosmer–Lemeshow test with $g = 10$ yields $H = 4.406$, $df = 8$, $p = 0.819$, so there is little evidence of miscalibration from this test. Also, the calibration plot looks adequate (not shown here). Therefore, we will accept this model as sufficiently well calibrated. The C-statistic is 0.753, indicating that the model discriminates moderately well between people who attend pop concerts and those who don't.

The primary research question was whether pop concert attendance is related to social status (measured by the z-standardised ISEI score), controlling for the other predictors in the model. We note, first of all, that the estimated odds ratio for ISEI is 1.11. This estimate is smaller than the analogous estimate from Model 2.5 predicting classical concert attendance, which was 1.21 (see the section 'An Example with Multiple Predictors'). This result is consistent with the idea that social status may have a larger influence on classical concert attendance than on pop concert attendance, although it is not clear whether the difference here is large enough for us to conclude that the data confirm our hypothesis. We may also wish to formally test whether there is evidence from Model 2.9 for a relationship between social status and pop concert attendance. An LRT comparing Model 2.9 to an identical model, but with the coefficient for ISEI set to zero, suggests that there is little evidence for such a relationship, when controlling for the other predictors ($\Lambda = 1.011$, $df = 1$, $p = 0.315$).

If a test of the effect of social status (ISEI) was our sole purpose in fitting this model, our interpretation won't necessarily focus on the other coefficients.[xxi] However, I will add a few observations, because they illustrate some interesting issues. Note the differences between the prediction of classical concert and pop concert attendance. The results of Model 2.5 were that older people, those with high school education, those with higher incomes and perhaps those with higher social status were more likely to attend classical concerts, controlling additionally for rural or urban residence and the presence or absence of young children in the household. (The latter two variables did not seem to have much of an influence on the probability of attending classical concerts.) In contrast, it is younger people who are more likely to attend pop concerts, and there is no strong evidence that either social status or education have a strong influence, controlling for the other variables in the model. On the other hand, having young children appears to

---

[xxi]However, even if our primary interest is only about some of the coefficients in the model, we should always check that all estimated coefficients make sense. This will decrease the risk of reporting results based on silly errors. For example, if our estimated coefficients had implied that people of pension age are more likely to attend pop concerts than 20-somethings, we may have wanted to check whether we have coded the age variable correctly.

make pop concert attendance slightly less likely, judging from this model. The evidence for rural residence and income is a bit unclear, since the confidence intervals include the 'no effect' odds ratio of 1, but nonetheless, the point estimates suggest the possibility of relatively strong effects.

An illustration similar to Figure 2.11 would show that age is by far the strongest predictor of pop concert attendance in this data set. To illustrate the size of the age effect, we might do one of several things. We might fix the values of all other predictors at some sensible values (e.g. to means and reference categories) and calculate predicted probability of pop concert attenders at various ages. For example, consider a person with lower than high school education (*Highschool* = 0), who lives in an urban area (*Rural* = 0) without young children (*Youngkids* = 0) and whose job has average social status (*z.ISEI* = 0). For such a person, Model 2.9 predicts their probability of attending pop concerts to be as follows:

$$p(Pop) = \frac{\exp\left(-0.316 - 0.085 \times AgeC_i - 0.0023 \times AgeC_i^2\right)}{1 + \exp\left(-0.316 - 0.085 \times AgeC_i - 0.0023 \times AgeC_i^2\right)}$$

We can use this equation to calculate example probabilities for various ages. For example, a 25-year-old with the characteristics mentioned above would have $p(Pop) = 0.71$. A 45-year-old with the same characteristics has $p(Pop) = 0.42$. And a 65-year-old has $p(Pop) = 0.03$. We thus see that age has a strong relationship with pop concert attendance, controlling for income, social status, rural/urban residence, presence/absence of young children and education.

Non-binary categorical variables

We have seen in this chapter how we can use logistic regression to analyse a binary outcome. But what to do if our outcome has three categories, or more? In that case, we cannot use binary logistic regression directly. We could do only so if we first simplified our outcome by dichotomising it – that is, by recoding it so that some categories are combined and we end up with two categories. However, dichotomisation results in loss of information and may make our analysis less relevant for our research question.

In many cases, a preferable option for the analysis of non-binary categorical outcomes is to use one of the methods introduced in the next two chapters: Chapter 3 deals with ordinal logistic regression, for the analysis of ordered categorical outcomes; Chapter 4 discusses multinomial logistic regression, for the analysis of nominal outcomes with three or more categories.

Chapter Summary

Binary outcomes are of interest in many areas of research, and so logistic regression is a commonly applied method in the social sciences, psychology, medical research, and other fields. Different scientific disciplines have different conventions of reporting some aspects of logistic regression. For example, the C-statistic is most commonly used in medical and epidemiological studies, whereas social science and psychological papers are more likely to report a pseudo-R^2 statistic instead.

Two alternatives to binary logistic regression should be mentioned:

- The *linear probability model* proposes a linear relationship between the probability of a binary outcome event and numeric predictors. As I suggested in the section 'Logistic Regression: The Model', the linear probability model has the disadvantage that it can lead to predicted probabilities that are impossible (larger than 1, or smaller than 0). This is why it is rarely used in the social sciences. However, the linear probability model does have the advantage that effect estimates may be more intuitive than the odds ratios one obtains from logistic regression. Using a linear probability model requires careful attention to some technical details. A good introduction and rationale for this model is provided by Battey et al. (2019).
- *Probit regression* aims to predict the probability of the binary outcome event in a different way than logistic regression. Probit regression uses the so-called probit transformation, which leads to predicted probabilities that are strictly in the interval between 0 and 1. Probit regression is a viable alternative to logistic regression, but it is less often used. Often the results of a logistic and probit regression on the same data will be very similar in terms of their substantively important interpretation.

Further Reading

Agresti, A. (2013). *Categorical data analysis* (3rd ed.). Wiley.

Hilbe, J. M. (2009). *Logistic regression models*. Chapman & Hall.

You could consult these books for more extensive and mathematically thorough descriptions of binary logistic regression.

Liao, T. F. (1994). *Interpreting probability models: Logit, probit, and other generalized linear models*. Sage.

Liao's book has a useful focus on the interpretation of logistic regression models and goes into more depth than the book you are reading now.

Long, J. S., & Freese, J. (2014). *Regression models for categorical dependent variables using Stata* (3rd ed.). Stata Press.
This book offers you detailed instruction on how to use logistic regression with the Stata software.

Tabachnick, B. G., & Fidell, L. S. (2014). *Using multivariate statistics* (6th ed.). Pearson.
The excellent textbook has a chapter on logistic regression with data analysis examples.

3

ORDINAL LOGISTIC REGRESSION: THE GENERALISED ORDERED LOGIT MODEL

Chapter Overview

The previous chapter considered logistic regression, a model for the analysis of a binary outcome. This chapter introduces models for the analysis of ordinal categorical outcomes. We call a variable 'ordinal' when its categories can be placed in a meaningful order. Ordinal variables are common in social research. Some examples are:

- *Highest qualification:* 'no qualification', 'primary school only', 'secondary school', or 'university degree or equivalent'
- *Subjective health status:* 'poor', 'fair', 'good', or 'very good'
- *Alcohol consumption:* 'non- or occasional drinkers', 'low-risk drinkers', 'moderate drinkers who occasionally binge drink', 'heavy drinkers', or 'problem drinkers' (Caldwell et al., 2008)
- Also, many survey questions designed to measure attitudes and opinions have ordinal response scales, such as 'strongly agree', 'agree', 'neither agree nor disagree', 'disagree', or 'strongly disagree'.

An example of a categorical variable that is *not* ordinal is 'political party voted for at the last election'. For example, in England, this variable might have the values 'Conservative Party', 'Labour Party', 'Liberal Democrats', 'Greens', and so forth. This is called a nominal variable. The analysis of nominal outcomes is considered in Chapter 4.

This chapter introduces the **generalised ordered logit model** – or **gologit model** for short. As its name implies, the gologit model employs a logit transformation of the outcome. It is thus a type of logistic regression for ordinal outcomes and is closely related to logistic regression for binary outcomes (see Chapter 2). In fact, another name for the gologit model is **ordinal logistic regression**. However, some authors use the term *ordinal logistic regression* to refer to a particular type of gologit model only, namely, the so-called proportional odds model. In contrast, this chapter distinguishes three types of gologit models, and thus three types of ordinal logistic regression: (1) the *proportional odds model*, (2) the *partial proportional odds model* and (3) the *non-proportional odds model*. To understand the differences between these three models, as well as their advantages and disadvantages, let's first look at a simple example of data that involve an ordinal outcome.

Modelling ordinal outcomes: proportional odds or non-proportional odds?

Consider Table 3.1. This shows data from the Faith Matters Survey, which was conducted in the USA in 2006. Lim and Putnam (2010) analysed these data to investigate theories about the relationship between religious belief, religious belonging, and life satisfaction. Previous studies had found that religious people in the USA tended to report higher life satisfaction than non-religious people. Lim and Putnam wanted to find out why. Their outcome, life satisfaction, was measured as an ordinal variable, whose categories

I have coded 'low', 'medium' or 'high' for the purpose of Table 3.1.[1] Respondents were also asked to state how important religion was to their lives. Responses were coded 'very important' or 'not very important', and the resulting variable was called 'religious identity'. Religious identity serves as a predictor in our analysis.

Table 3.1 Numbers of people who report low, medium and high life satisfaction in the US Faith Matters Survey, 2006, by religious identity

Religious Identity	Life Satisfaction		
	Low	Medium	High
Not very important	160	400	334
Very important	132	507	694

Cutpoint 1   Cutpoint 2

Note. The data for this chapter were taken from the Faith Matters Survey, 2006. The principal investigators of the survey were Robert D. Putnam and David E. Campbell. The data were downloaded from the Association of Religion Data Archives (www.thearda.com/Archive/Files/Descriptions/FTHMATT.asp).

How can we model the relationship between religious identity and life satisfaction? Since the outcome, life satisfaction, has three categories, we cannot apply binary logistic regression, at least not directly. However, we could turn our three-category variable into a binary variable in either of two ways:

- We could reclassify respondents into those with 'low life satisfaction' versus those with 'medium or high life satisfaction', thus dichotomising our outcome by making a cut between 'low' and 'medium'. Let's call this cutpoint 1.
- Or we could reclassify respondents into those with 'low or medium life satisfaction' versus 'high life satisfaction', making a cut between 'medium' and 'high'. Let's call this cutpoint 2.

These two dichotomisations of the life satisfaction variable are shown as two separate 2 × 2 tables in Table 3.2.

Table 3.2 Two ways of dichotomising the life satisfaction data from Table 3.1

Religious Identity	Life Satisfaction: Cutpoint 1		Life Satisfaction: Cutpoint 2	
	Low	Medium or High	Low or Medium	High
Not very important	160	734	560	334
Very important	132	1201	639	694

---

[1]The life satisfaction variable actually had 10 response categories in the survey (from 1 = 'extremely dissatisfied' to 10 = 'extremely satisfied'). I have recoded these into three categories to make the example simpler to explain. In principle, it is possible and indeed appropriate to use all 10 categories in an ordinal logistic regression model.

Now, for each 2 × 2 table, we may calculate the ratio of the odds of being in the more satisfied category, comparing those with a strong religious identity ('very important') to those without. For cutpoint 1, this gives

$$\widehat{OR}_1 = \frac{132/734}{160/1201} = 1.98$$

For cutpoint 2, we have

$$\widehat{OR}_2 = \frac{560/639}{334/694} = 1.82$$

Thus, in this data set, people with strong religious identity have 1.98 times higher odds of reporting medium or high life satisfaction, rather than low life satisfaction, compared to people without a strong religious identity (cutpoint 1). Also, those with a strong religious identity have 1.82 times higher odds of reporting high life satisfaction, rather than low or medium life satisfaction (cutpoint 2). Note that the two odds ratios, although not exactly the same, are similar.

We can now set up a generalised ordered logit model in two different ways:

1   *Proportional odds model*: We could make the assumption that the two odds ratios (from cutpoints 1 and 2) are the same in the population from which the data were drawn. This is called the proportional odds assumption. The two odds ratios we observed (1.98 and 1.82) would then be considered as two estimates of the same population odds ratio. To reflect this assumption, the proportional odds model is defined so that it yields just one estimate of the population odds ratio.
2   *Non-proportional odds model:* Alternatively, we could assume that the two odds ratios are different in the population. This would lead us to choose a model that estimates the two odds ratios separately. This type of model is called a 'non-proportional odds' model.

Examples of data that feature either exactly proportional or clearly non-proportional odds are given in Box 3.1.

Box 3.1

Illustration of Proportional Odds and Non-Proportional Odds

Consider two simple examples of men's and women's responses to some attitude question with three response categories. Table 3.3 gives an example of exactly proportional odds. For each of the two cutpoints, men's odds of being in the group that agrees more are twice as high as women's odds, so the odds ratio is 2 for each cutpoint. An interpretation of the results in Table 3.3 might be that men are more likely to agree and less likely to disagree than women.

Table 3.3 A hypothetical example of exactly proportional odds

	Strongly Disagree	Partly Agree, Partly Disagree		Strongly Agree
Men	200	200		200
Women	300	180		120
Cutpoint		>Strongly disagree	>Partly agree, partly disagree	
Odds (men)		$\dfrac{200 + 200}{200} = 2$	$\dfrac{200}{200 + 200} = 0.5$	
Odds (women)		$\dfrac{180 + 120}{300} = 1$	$\dfrac{120}{300 + 180} = 0.25$	
Odds ratio (men vs women)		$\dfrac{2}{1} = 2$	$\dfrac{0.5}{0.25} = 2$	

In contrast, the data in Table 3.4 indicate that the odds are not proportional. Men's odds of being in a category above 'strongly disagree' are greater than women's, so the odds ratio for cutpoint 1 is greater than 1. On the other hand, men's odds of being in the 'strongly agree' category are lower than women's, so the odds ratio for cutpoint 2 is smaller than 1. An interpretation of Table 3.4 might be that women are more likely than men to have a strong opinion (either in agreement or in disagreement), whereas men are more likely than women to sit on the fence (i.e. to partly agree and partly disagree).

Table 3.4 A hypothetical example of clearly non-proportional odds

	Strongly Disagree	Partly Agree, Partly Disagree		Strongly Agree
Men	200	200		200
Women	240	120		240
Cutpoint		>Strongly disagree	>Partly agree, partly disagree	
Odds (men)		$\dfrac{200 + 200}{200} = 2$	$\dfrac{200}{200 + 200} = 0.5$	
Odds (women)		$\dfrac{120 + 240}{240} = 1.5$	$\dfrac{240}{240 + 120} = 0.67$	
Odds ratio (men vs women)		$\dfrac{2}{1.5} = 1.33$	$\dfrac{0.5}{0.67} = 0.75$	

Let's see how our two ordered logistic models would look for the data in Table 3.1. We wish to predict life satisfaction by religious identity.[ii] Let's define Religious Identity (RELID) as a dummy variable coded 1 = very important and 0 = not very important.

We set up a proportional odds model as follows.

Model 3.1: Proportional odds model

$$\text{logit}\left[P(LIFESAT_i > \text{'low'})\right] = \alpha_1 + \beta_1\,RELID_i$$

$$\text{logit}\left[P(LIFESAT_i > \text{'medium'})\right] = \alpha_2 + \beta_1\,RELID_i$$

The model consists of two equations. Each equation on its own looks similar to a binary logistic regression. The first equation estimates the log odds of having higher than low life satisfaction. This corresponds to cutpoint 1 (see Table 3.1). The second equation estimates the log odds of having higher than medium life satisfaction (i.e. of having high life satisfaction; this corresponds to cutpoint 2).

Note that β_1, the coefficient of RELID, appears in both equations. This reflects the proportional odds assumption. As in binary logistic regression, $\exp(\beta_1)$ is the odds ratio associated with a 1-point difference in the predictor. Model 3.1 assumes that the coefficient β_1 and therefore the odds ratio, $\exp(\beta_1)$, are the same for both cutpoints.

Note an important difference to binary logistic regression. In binary logistic regression, we had only one model equation, and this had a single intercept (β_0). The proportional odds model instead has several equations, and each equation features a different intercept. In this simple example, we have two equations, and for each there is a separate intercept, denoted here by α_1 and α_2.

Usually, when interpreting a model, we are more interested in the slope coefficients than in the intercepts. But it's useful to understand what the intercepts mean. The intercepts represent the baseline log odds of being in the higher outcome category for each dichotomisation. Baseline log odds here means 'the log odds for those who have the value zero on all predictor variables'. In our simple case, the intercepts give the log odds for the reference group of RELID – that is, those who report that religion is 'not very important' to them. So α_1 represents the log odds of having higher than 'low' life satisfaction for the reference group, while α_2 represents the log odds of having higher than 'medium' life satisfaction for this group.

Now, let's contrast the proportional odds model with the non-proportional odds model. A non-proportional odds gologit model looks like this:

[ii] A person's life satisfaction has multiple sources and causes, so I don't propose that religious identity is the only predictor of life satisfaction – but let's consider this simplistic model for the sake of an example.

Model 3.2: Non-proportional odds model

$$\text{logit}\left[P(LIFESAT_i > \text{'low'})\right] = \alpha_1 + \beta_{1,1}\, RELID_i$$

$$\text{logit}\left[P(LIFESAT_i > \text{'medium'})\right] = \alpha_2 + \beta_{1,2}\, RELID_i$$

In contrast to the proportional odds model (Model 3.1), the non-proportional odds model (Model 3.2) has a different slope coefficient for the predictor in each equation. We call these coefficients $\beta_{1,1}$ and $\beta_{1,2}$, to denote the coefficients for predictor 1 in the first equation, and for predictor 1 in the second equation, respectively. Model 3.2 does *not* assume these coefficients to be equal, and hence also doesn't assume the corresponding odds ratios, $\exp(\beta_{1,1})$ and $\exp(\beta_{1,2})$, to be the same.

To illustrate the use of the proportional odds and non-proportional odds models, let's estimate Models 3.1 and 3.2 on the data from the Faith Matters Survey. The results are displayed in Table 3.5.

Table 3.5 Estimates from two ordinal logistic regression models of life satisfaction on religious identity

	Estimate	SE	Odds Ratio: exp(β)
a. Proportional odds model (Model 3.1)			
RELID: very important (β_1)	0.62	0.08	1.86
Intercepts:			
LIFESAT > low (α_1)	1.55	0.08	
LIFESAT > medium (α_2)	−0.53	0.07	
b. Non-proportional odds model (Model 3.2)			
(1) *LIFESAT* > low			
RELID: very important ($\beta_{1,1}$)	0.68	0.13	1.98
Intercept (α_1)	1.52	0.09	
(2) *LIFESAT* > medium			
RELID: very important ($\beta_{1,2}$)	0.60	0.09	1.82
Intercept (α_2)	−0.52	0.07	

Note. RELID = religious identity; *LIFESAT* = life satisfaction; *SE* = standard error.

The estimated model equations for the proportional odds model (Model 3.1) are as follows:

$$\text{logit}\left[P(LIFESAT_i > \text{'low'})\right] = 1.55 + 0.62 \times RELID_i$$

$$\text{logit}\left[P(LIFESAT_i > \text{'medium'})\right] = -0.53 + 0.62 \times RELID_i$$

And the estimated equations for the non-proportional odds model (Model 3.2) are as follows:

$$\text{logit}\left[P(LIFESAT_i > \text{'low'})\right] = 1.52 + 0.68 \times RELID_i$$

$$\text{logit}\left[P(LIFESAT_i > \text{'medium'})\right] = -0.52 + 0.60 \times RELID_i$$

So we see that the proportional odds model estimates only one slope coefficient for RELID, whereas the non-proportional odds model estimates two slope coefficients, one for each cutpoint. As in binary logistic regression, the odds ratio provides an estimate of the direction and strength of the relationship between the predictor and the outcome. For the proportional odds model, the estimated odds ratio is 1.86. Thus, the model estimates that those with strong religious identity have 86% higher odds of being in a higher life satisfaction category, compared to those for whom religion is not very important. This model assumes the proportional odds assumption to be valid.

For the non-proportional odds model, we have two odds ratios to interpret. The odds ratio of 1.98 for the first cutpoint indicates that those people with a strong religious identity have 1.98 times higher odds than those without a strong religious identity of being in a life satisfaction category higher than 'low', according to this model. The odds ratio of 1.82 for the second cutpoint indicates that those with a strong religious identity have 1.82 times higher odds of having 'high' life satisfaction, according to this model.

It might seem to you that the non-proportional odds model is better, or 'safer', since it does not rely on the assumption of proportional odds to be valid. However, if it is true that the odds ratios are indeed the same in the population, then the proportional odds model has advantages: because it uses all the data to estimate a common coefficient for each predictor, this coefficient will usually have a smaller standard error than any of the several coefficients that would be estimated by a corresponding non-proportional odds model. Also, having a single odds ratio per predictor, rather than several, will often make the results of a proportional odds model easier to interpret than those of a corresponding non-proportional odds model.

How do we choose between these two models? The answer is that we can use likelihood ratio tests and other procedures to investigate whether the proportional odds assumption is plausible – and if it is, then it is usually a good idea to choose the proportional odds model. However, in general, our models will have more than one predictor, and so we will face the choice between assuming and not assuming proportional odds for *each* predictor variable. This will lead us to a third type of gologit model, the partial proportional odds model, which assumes proportional odds for some predictors, but not all. Before we discuss in detail how to choose between the three different types of gologit model in a given research situation, let's look at how we can further illustrate the results from any ordinal logistic regression model.

Calculating predicted probabilities

In the interpretation of an ordinal logistic regression, it often helps to consider how the predicted probability of being in each category of the outcome varies with the predictor, or predictors. I will demonstrate this below for the proportional odds model (Model 3.1), but the method would work analogously for the non-proportional odds model too.

We can rewrite the equations from an ordinal logistic regression so that we obtain predicted probabilities (just as we can in binary logistic regression). To illustrate, let's do this for the proportional odds model, Model 3.1 (see Table 3.5). The predicted probability of having medium or higher life satisfaction according to the estimates from Model 3.1 is as follows:

$$P\left(LIFESAT_i > 'low'\right) = \frac{\exp\left(\alpha_1 + \beta_1\,\mathrm{RELID}_i\right)}{1 + \exp\left(\alpha_1 + \beta_1\,\mathrm{RELID}_i\right)} = \frac{\exp\left(1.55 + 0.62 \times RELID_i\right)}{1 + \exp\left(1.55 + 0.62 \times RELID_i\right)}$$

Doing this calculation gives $P(LIFESAT_i > 'low') = 0.83$ for those without a strong religious identity and $P(LIFESAT_i > 'low) = 0.90$ for those with a strong religious identity.

The analogous calculation for the second equation from Model 3.1 gives us the probabilities to have high life satisfaction. These are $P(LIFESAT_i > 'medium) = 0.37$ for those without a strong religious identity and $(LIFESAT_i > 'medium) = 0.52$ for those with a strong religious identity.

From these two sets of predicted probabilities, we can derive predictions for each of the three categories of life satisfaction, since

$$P\left(LIFESAT_i = 'low'\right) = 1 - P\left(LIFESAT_i > 'low'\right) \text{ and}$$
$$P\left(LIFESAT_i = 'medium'\right) = P\left(LIFESAT_i > 'low'\right) - P\left(LIFESAT_i > 'medium'\right)$$

The predicted probabilities are shown in Figure 3.1. This illustrates the result that a strong religious identity is associated with higher life satisfaction in this data set, according to Model 3.1.

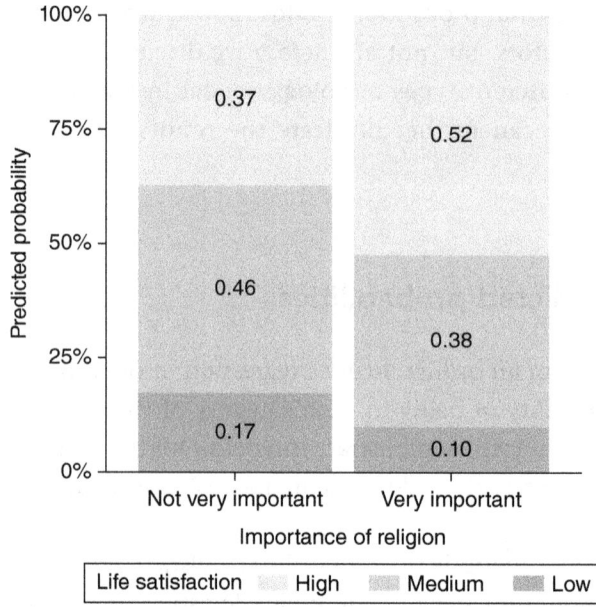

Figure 3.1 Predicted probabilities of low, medium and high life satisfaction from a proportional odds ordinal regression on subjective importance of religion (from Model 3.1)

The proportional odds ordinal logistic regression model

We will now look at the formal definition of the proportional odds ordinal logistic regression model. In the examples we have considered so far, we had one predictor variable and three outcome categories. This section defines the model for any number of outcome categories and any number of predictors. Let's assume that we have an ordinal outcome, Y, with categories $j = 1, 2, \ldots, J$. Let p_j be the probability of being in a category higher than j. That is, $p_j \equiv P(Y_i > j)$, where the symbol $\equiv$ means 'is defined as'.

The proportional odds ordinal logistic regression model is

$$\text{logit}(p_j) = \log\left(\frac{p_j}{1 - p_j}\right) = \alpha_j + \beta_1 x_{1i} + \beta_2 x_{2i} + \ldots + \beta_k x_{ki}, \quad j = 1, \ldots, j - 1,$$

where the ordinal response categories of the outcome are numbered $j = 1, 2, \ldots, J$.[iii]

[iii]This way of writing the model differs from some other authors'. Whereas the model presented here predicts the probability of being in a higher category, many authors formulate the model the other way around, choosing instead to predict the probability of being in a lower category. This leads to the same results in terms of odds ratios and predicted probabilities, but has the disadvantage that the slope coefficients have a negative sign in the model equation, which seems needlessly confusing. See also Long and Freese (2014). Yet another variant of writing this model is with negative signs for the intercepts. Therefore, when interpreting output from estimating an ordinal logistic regression model in statistical software, it is important to make sure you know how the software defines the model equation and its coefficients – otherwise you may misinterpret the coefficient estimates in the output.

We can write this model using $J - 1$ separate equations – that is to say, we always have one fewer equation than we have outcome categories. For example, Model 3.1 consists of two equations, because the outcome has three categories (see the section 'Modelling Ordinal Outcomes: Proportional Odds or Non-Proportional Odds?'). Each equation has its own intercept α_j, but in a proportional odds model the slope coefficients $\beta_1, \beta_2, \ldots, \beta_k$ are the same across all equations.

Ordinal logistic regression is similar to binary logistic regression in many ways. Like in binary logistic regression, the coefficients in an ordinal logistic regression model are found via the method of maximum likelihood. Consequently, we can

- use the likelihood ratio test to compare nested models and
- calculate confidence intervals by the profile likelihood method.

Just like in binary logistic regression, hypothesis tests and confidence intervals based on the normal approximation are often displayed by default in statistical software. These tests and confidence intervals will often agree with their likelihood-based counterparts, but if they do not, the likelihood ratio test and profile likelihood confidence intervals should be preferred.

Two crucial differences between binary and ordinal logistic regression are:

- There is no equivalent for the Hosmer–Lemeshow test for ordinal logistic regression. Thus, model calibration needs to be investigated by graphical means only.
- In an ordinal logistic model, we have to decide whether we wish to assume proportional odds or not. We can investigate the proportional odds assumption using a statistical test called Brant's test (see the section 'Testing the Proportional Odds Assumption: Brant's Test').

We will now look at a small example of a proportional odds model with two predictors to illustrate these features. Let's consider the question how two demographic variables, age and gender, relate to life satisfaction. In the Faith Matters Survey data set, age is a continuous variable ranging from 18 to 95 years, with a median age of 50 and a mean age of 50.5. Gender is a binary variable; 56% of respondents were female, 44% were male.

To explore the shape of the relationship between age and life satisfaction, I have categorised age and plotted it against life satisfaction in Figure 3.2.

Figure 3.2 suggests that there may be a relationship between age and life satisfaction, although this relationship does not seem to be linear. People older than 60 years in this sample are most likely to report high life satisfaction. The middle-aged (40–59 years) are least likely to be highly satisfied. People in their 30s seem to be more likely to be satisfied than the middle-aged, but not as much as the over-60s. The life satisfaction ratings of people under 30 years look similar to those of the middle-aged.

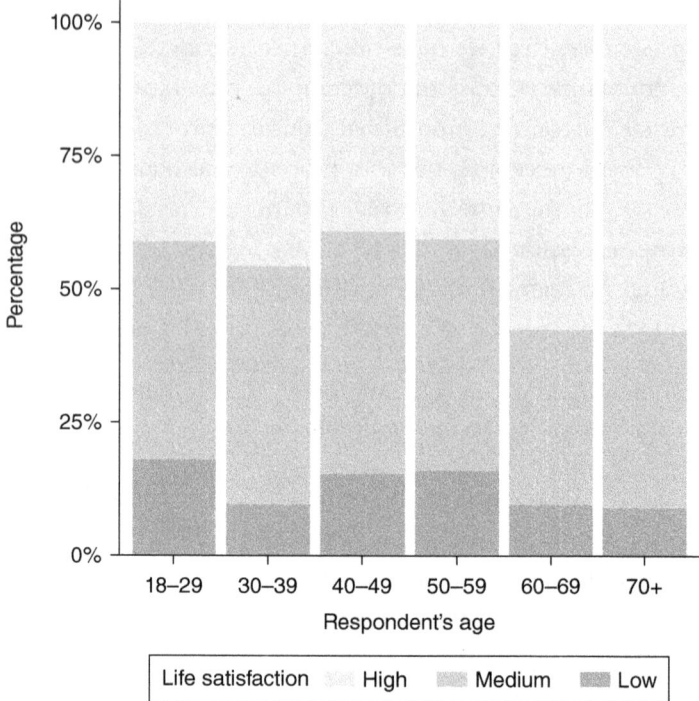

Figure 3.2 The distribution of life satisfaction in the Faith Matters Survey, 2006, by age category

Thus, the relationship between age and life satisfaction may be complex. But we should also remember that the results in this sample don't necessarily exactly reflect the true relationship in the population. We aim for a statistical model that is as complex as necessary to not be evidently contradicted by the data, but that is otherwise as simple as possible.

In this example, let's consider two hypotheses about the relationship between age and life satisfaction:

- The first hypothesis posits that the older a person we ask, the more likely they are to report a higher level of life satisfaction. This hypothesis, if true, would suggest that we might model the relationship between life satisfaction and age by a single linear term for age, such that we can assume linearity in the relationship between age and the log odds of being in a higher life satisfaction category.
- The second hypothesis states that life satisfaction, on average, starts at a reasonably high level in the first two decades of adulthood, then tends to decline somewhat during middle age, and increase to its highest levels in older age

(among those who live long enough to experience it). This hypothesis thus suggests a U-shaped relationship between age and life satisfaction, which would imply that we should include a quadratic term for age in our model.

Of course, other hypotheses are possible. To keep the example simple, let's investigate just these two. We test two models against one another: Model 3.3 attempts to predict life satisfaction by gender and age. Gender is represented by a dummy variable called *Male*, coded 0 = *female* and 1 = *male*. AgeC is a continuous variable that represents age centred on its median (such that 0 indicates 50 years of age) and measured in decades. Model 3.4 is identical to Model 3.3, except that we add a term for the square of the centred age variable. So we have the following:

Model 3.3: Linear age effect

$$\text{logit}\left[P(LIFESAT_i > j)\right] = \alpha_j + \beta_1 Male_i + \beta_2 AgeC_i$$

Model 3.4: Linear + quadratic age effect

$$\text{logit}\left[P(LIFESAT_i > j)\right] = \alpha_j + \beta_1 Male_i + \beta_2 AgeC_i + \beta_3 AgeC_i^2,$$

where

- $LIFESAT_i$ is the life satisfaction reported by the ith respondent (low, medium or high);
- $j = 1, 2$, where 1 = low and 2 = medium;
- $AgeC$ is continuous age centred on its median and then divided by 10 ($AgeC = Age - 50)/10$, so that the median age is zero and age is measured in decades[iv];
- $AgeC^2$ is the square of $AgeC$; and
- $Male$ is a dummy variable with reference category 'Female'.

Since Model 3.3 is nested within Model 3.4, we can conduct a likelihood ratio test to compare the two models. The null hypothesis is that Model 3.4 explains the variation

---

[iv]Why did I decide to measure age in decades, rather than years? It is simply a matter of convenience. The estimated coefficients for the effect of 1 year of age, and especially for the effect of 1 unit of age-in-years squared, would be very small, and so would their standard errors be. Thus, I would need to display many decimal points to show them with reasonable precision. By measuring age in decades, I obtain an estimate of the effect of a 10-year difference in age on the log odds of being in a higher life satisfaction category. Changing the scale on which I measure age in this way does not affect the substantial interpretation of my results. Intercepts and coefficients of other covariates in the model are not affected by the choice of scale for age, and neither are predicted probabilities.

in life satisfaction no better than Model 3.3. The alternative hypothesis is that Model 3.4 provides a better prediction of life satisfaction than Model 3.3. The result of this test is displayed in Table 3.6.

Table 3.6 Likelihood ratio test of Model 3.4 versus Model 3.3

		Log Likelihood	Λ	Parameters	df	df Difference	p
Model 3.3	Male + AgeC	−2182.49		4	2223		
Model 3.4	Male + AgeC + AgeC²	−2177.54	9.89	5	2222	1	0.0017

Note. AgeC = (Age − 50)/10, so that age is centred on 50 years and measured in decades; df = degrees of freedom.

The likelihood ratio test statistic is 9.89, with 1 *df*. This yields a *p*-value of 0.0017. Thus, there is strong statistical evidence against the null hypothesis. This finding indicates that we should prefer Model 3.4 over Model 3.3, which implies that the relationship between age and the log odds of being in a higher life satisfaction category is modelled better by a combination of a linear and quadratic term than by a linear term alone. The estimated coefficients from Model 3.4 are shown in Table 3.7.

Table 3.7 Estimates from the ordinal logistic regression of life satisfaction on gender and age (Model 3.4)

	Estimate	SE	OR	95% CI for OR
Male	−0.243	0.082	0.78	[0.67, 0.92]
AgeC (in decades)	0.123	0.015	1.13	[1.08, 1.19]
AgeC²	0.041	0.013	1.04	[1.02, 1.07]
Cutpoints				
LIFESAT > low (α_1)	1.907	0.081		
LIFESAT > medium (α_2)	−0.169	0.066		

Note. 95% confidence intervals were calculated using the profile likelihood method. AgeC = (Age − 50)/10, so that age is centred on 50 years and measured in decades. LIFESAT = life satisfaction; SE = standard error; OR = odds ratio; CI = confidence interval. Data were taken from the Faith Matters Survey, 2006.

The negative coefficient for *Male* indicates that, controlling for age and age squared, males tend to report lower life satisfaction than females. The odds ratio and its confidence interval for *Male* suggests that a man has between 8% and 33% lower odds of being in a higher life satisfaction category, compared to a woman of the same age.

The odds ratios for *AgeC* and *AgeC²* need to be interpreted together (since I can't change one while holding the other constant). To illustrate the effect of a continuous predictor in a gologit model, it is often useful to draw a graph that shows how the predicted probabilities of the outcome categories depend on the predictor. For Model 3.4, this is shown in Figure 3.3.[v]

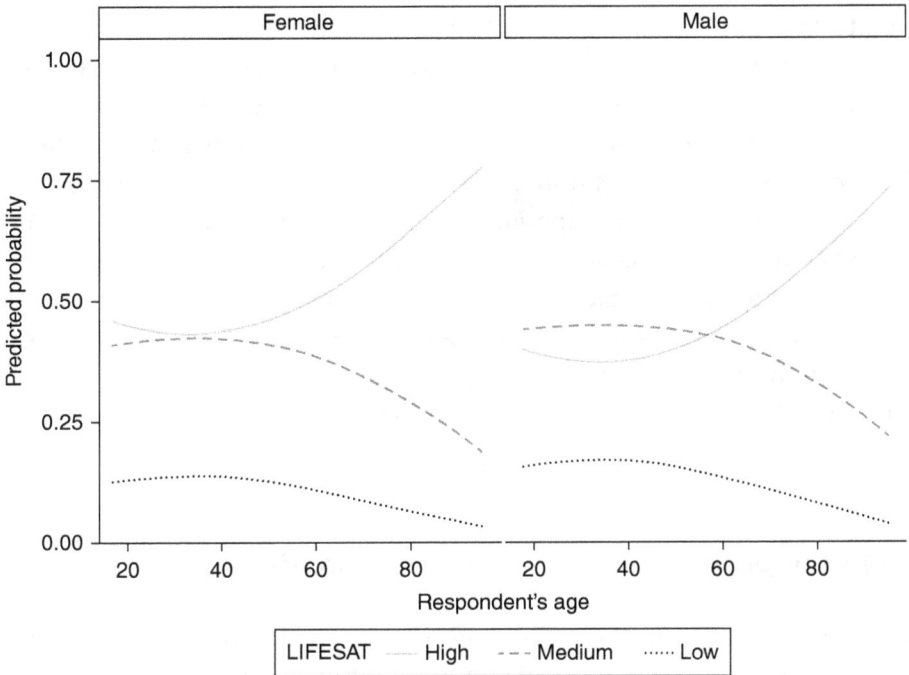

Figure 3.3 The relationship between life satisfaction, age and gender, predicted by Model 3.4

There are gender differences in life satisfaction, such that women tend to report higher life satisfaction than men at all ages. However, the variations by age are far larger than the differences between the genders. We see that adding both a linear and a quadratic term for age has resulted in a predicted relationship similar to what we saw in the observed proportions shown in Figure 3.1.

Why are life satisfaction and age related in this way? That is a question that neither these data nor the model can help us answer by themselves. We need to consider how the data were collected, and what the likely social and biological mechanisms are that might lead to this relationship. An example of a consideration about the data collection process is to reflect that this is a cross-sectional study (see *The SAGE Quantitative Research Kit*, Volume 1), not a longitudinal one. We have not followed the same people over time, so we have not observed them as they have aged. Thus, we cannot conclude that 'as people get older beyond the age of

---

[v]We can derive predicted probabilities by the same method as shown above in the section 'Calculating Predicted Probabilities'. I will not show the calculations here.

50 years, their life satisfaction increases'. That would be incorrect, because survivorship bias is likely to play a role in our results. Consider that not all the middle-aged people in this sample will live to see old age and that life satisfaction may well be related to life expectancy. For example, people with a severe illness may be less likely to be satisfied with their lives, and may also be at higher risk of dying before the age of 60 than people without such an illness. Thus, our sample of 60-year-olds is not a random sample of the 40-year-olds of 20 years ago. It may be that the more satisfied of those 40-year-olds from 20 years ago had a higher chance of surviving to age 60. This is an example of (potential) survivorship bias. Thus, the relationship between age and life satisfaction that we observe in our data, and that we describe using our statistical model, may in part be an artefact of our cross-sectional research design. Of course, it is possible that, at the same time, there are other reasons why life satisfaction should be related to age, and that these reasons represent real 'age effects'. I won't discuss this question further here.

Testing the proportional odds assumption: Brant's test

Before we interpret the results from our proportional odds model further, we should investigate whether the proportional odds assumption holds. A statistical test called Brant's test is designed to investigate this assumption. The test works as follows: first one estimates the binary logistic regressions that correspond to all possible dichotomisations of the outcome. Then one tests the null hypothesis that the resulting observed coefficients are all estimates of the same underlying common coefficient. For example, for Model 3.4, two binary logistic regressions would be calculated: one that predicts high life satisfaction (vs low or medium) and another that predicts having at least medium satisfaction (vs low). This would result in two different estimated odds ratios for each of the predictors. Brant's test investigates whether these observed binary odds ratios are similar enough to consider them as estimates of the same population odds ratio.

There are two commonly used versions of Brant's test:

1   Brant's omnibus test investigates equality of coefficients for all predictors jointly. The null hypothesis of Brant's omnibus test is that the odds are proportional for all predictors in the model. The alternative hypothesis is that the odds are non-proportional for at least one predictor.

2   Brant's test for an individual predictor investigates equality of coefficients for one specific predictor. The null hypothesis here is that the odds are proportional for this predictor, controlling for the other variables in the model. The alternative hypothesis is that the odds for this predictor are not proportional.

Under the null hypothesis, either version of Brant's test statistic has a χ^2 (chi-square) distribution. For a test of an individual predictor, the degrees of freedom equal the number of outcome categories minus two. The omnibus test has degrees of freedom equal to the sum of the degrees of freedom of the individual predictors. Results of Brant's test for Model 3.4 are displayed in Table 3.8.

Table 3.8 Results of Brant's tests and Brant's omnibus test for the coefficients of Model 3.4

	Test Statistic (Chi-Square)	df	p
Omnibus	2.26	3	0.52
Male	0.91	1	0.34
AgeC	0.11	1	0.74
AgeC²	1.17	1	0.28

Note. AgeC = continuous age centred on 50 and measured in decades; *AgeC²* = continuous age squared; *df* = degrees of freedom.

I recommend a principled strategy in interpreting these results: first, consider only the result of Brant's omnibus test. If this has a large *p*-value (larger than 0.05, say), we may be justified in retaining our trust in the proportional odds assumption for all predictors. In that case, we don't need to consider the Brant's tests for individual predictors (even though some software packages will, by default, display them anyway, as in Table 3.8). The reason for looking at the omnibus test first is to decrease the risk of type I error inflation due to multiple testing.[vi]

On the other hand, if Brant's omnibus test yielded a small *p*-value, this would suggest that the odds are non-proportional for at least one of our predictors. In that case, it may be useful to conduct Brant's tests for the individual predictors, in order to find out where the problem lies – that is, for which predictor the proportional odds assumption appears to be violated.

In Table 3.8, the *p*-value for the omnibus test is quite large (and certainly larger than any conventional significance level). So there is no strong evidence against the

---

[vi]For example, consider a proportional odds model with 20 predictors. If the proportional odds assumption holds for all 20 predictors, and I nonetheless conduct 20 individual Brant's tests, on average I would expect to find one result with $p < 0.05$. So my risk of rejecting the proportional odds assumption for at least one predictor would be high, even though the assumption in fact holds. More generally, the more predictors I have, and the more Brant's tests I conduct, the higher is the risk that I reject the proportional odds assumption for some predictors, even if it is true for all. This risk is reduced if I consider the result of Brant's omnibus test first and only conduct the Brant's tests for individual predictors if the omnibus test indicates a violation of the proportional odds assumption.

proportional odds assumption. This suggests that we can trust the common odds ratios estimated by Model 3.4.[vii]

If the proportional odds assumption does not hold for the relationship between the outcome and one or several predictors, then the estimated coefficients from a proportional odds model may be misleading for *all* predictors (not just those for whom the proportional odds assumptions is violated). If you have reason to think that the odds for one or several predictors are not proportional, the following strategies may be tried:

- Transform the predictor, or predictors, for whom the odds are not proportional
- Add further meaningful predictors to your model.

If none of these helps to make the proportional odds assumption plausible, you may consider a different gologit model such as a non-proportional model or a partial proportional odds model. We will consider these models next.

Generalised ordinal logit models: full, partial and non-proportional odds

This section revisits the non-proportional odds model, which we first discussed in the section 'Modelling Ordinal Outcomes: Proportional Odds or Non-Proportional Odds?', and introduces it more formally. This sets the foundation for the partial proportional odds model, which assumes proportional odds for some predictors but not for the others. By the end of this section, you will have been introduced to three types of models: proportional odds, non-proportional odds and partial proportional odds. These are all variants of the generalised ordered logit model. We begin with an example of a relationship that fails to satisfy the proportional odds assumption.

A case where the proportional odds assumption is not met

Let's now consider a situation where the proportional odds assumption does not hold for the relationship between an ordinal outcome and a predictor variable. In the Faith Matters Survey, respondents were asked how many close friends they had and

---

[vii]Even though in this case we don't need to consider the Brant's test for individual predictors, Table 3.8 displays them for the purpose of illustration. In this instance, all the Brant's tests for individual predictors also yield large *p*-values, and thus the results are consistent with the omnibus test.

were given the response options: 'none', '1–2', '3–5', '6–10' and '11 or more'. This variable is measured on an ordinal scale. To make it easier to work with as a predictor, Lim and Putnam (2010) turned it into a numeric variable by taking the number of friends to be the midpoint of each of the ranges given and coding '11 or more' as 11. They then log transformed the resulting numeric variable, as shown in Table 3.9. It is the log-transformed variable which they use in their statistical models. We will call this variable *LOGFRDS*.

Table 3.9 Response option, numeric recoding and log-transformation of 'How many close friends do you have?'

Categorical Response Option	Numeric Approximation: Number of Friends	LOGFRDS = log(1 + Number of Friends)	Frequency (%)
None	0	0	62 (2.8)
1–2	1.5	0.916	401 (18.0)
3–5	4	1.609	939 (42.2)
6–10	8	1.946	501 (22.5)
11 or more	11	2.485	324 (14.5)

Note. N = 2227. Data from the Faith Matters Survey, 2006.

Now, imagine that we wish to investigate the relationship between life satisfaction and the number of friends. Having more friends may contribute to happiness and satisfaction. On the other hand, being happy with one's life may make it easier for some people to make new friends and to nurture existing friendships. Our data won't let us decide which direction the causation goes, but let's nonetheless investigate the relationship between life satisfaction and *LOGFRDS*, controlling for age and gender. I first propose a proportional odds model:

Model 3.5: Proportional odds model

$$\text{logit}\left[P\left(LIFESAT_i > j\right)\right] = \alpha_j + \beta_1 Male_i + \beta_2 AgeC_i + \beta_3 AgeC_i^2 + \beta_4 LOGFRDS$$

Since this is a proportional odds model, we have a single coefficient for each predictor, including for *LOGFRDS*. We estimate the model (estimated coefficients are shown further below and we will look at them later) and conduct Brant's omnibus test. The results are shown in Table 3.10. We see that Brant's omnibus test yields a small p-value ($p = 0.032$). Thus, we have reason to think that the proportional odds assumption may fail to hold for at least one predictor. Consequently, we conduct a Brant's test for each individual predictor. The tests for Male, AgeC and AgeC² all yield large p-values, suggesting that for these three predictors there is little evidence against the proportional odds assumption. However, the p-value of Brant's test for *LOGFRDS* is very small ($p = 0.004$), suggesting that the proportional odds assumption may not hold for this variable.

Table 3.10 Results of Brant's tests for Model 3.5

	Test Statistic (Chi-Square)	df	p
Omnibus	10.52	4	0.032
Male	1.08	1	0.298
AgeC	0.36	1	0.551
AgeC²	1.20	1	0.273
LOGFRDS	8.15	1	0.004

Note. AgeC = continuous age centred on 50 and measured in decades; AgeC² = continuous age squared; LOGFRDS = log(1 + friends); df = degrees of freedom.

As we saw in the previous section, there are now several ways in which we could try to address the problem, such as trying a different transformation of 'number of friends', or even modelling 'number of friends' as a categorical predictor (via four dummy variables). However, in this case, all alternatives I have tried have also resulted in a violation of the proportional odds assumption (I won't show these investigations here). So let's instead relax the proportional odds assumption.

The non-proportional odds model

The non-proportional odds model allows us to predict an ordinal outcome variable without assuming proportional odds for any predictor. The non-proportional odds model is defined as follows:

$$\text{logit}\left(p_j\right) = \log\left(\frac{p_j}{1-p_j}\right) = \alpha_j + \beta_{1j}x_{1i} + \beta_{2j}x_{2i} + \ldots + \beta_{kj}x_{ki}, \quad j = 1,\ldots,j-1$$

where the ordinal response categories of the outcome are numbered $j = 1, 2, \ldots, J$. In this model, each predictor variable has not one but several coefficients. This is represented in the mathematical notation by the subscript j for the slope coefficients (β_{1j}, β_{2j} and so forth). For example, the non-proportional model for the set of predictors we investigated in the previous section is as follows:

Model 3.6: Non-proportional odds model

$$\text{logit}\left[P\left(LIFESAT_i >' \text{low}'\right)\right] = \alpha_1 + \beta_{11}Male_i + \beta_{21}AgeC_i + \beta_{31}AgeC_i^2 + \beta_{41}LOGFRDS$$

$$\text{logit}\left[P\left(LIFESAT_i >' \text{medium}'\right)\right] = \alpha_2 + \beta_{12}MALE_i + \beta_{22}AGE.C_i + \beta_{32}AGE.C_i^2 + \beta_{42}LOGFRDS$$

Each predictor has two associated slope coefficients. For example, the coefficients of *LOGFRDS* are called β_{41} and β_{42}. The first, β_{41}, represents the effect of the number of friends on the log odds of having higher than 'low' life satisfaction. The second, β_{42}, represents the effect of number of friends on the log odds of having higher than 'medium' life satisfaction. Estimates from fitting this model to the Faith Matters Survey data are shown further below, and we will look at them later.

A disadvantage of the non-proportional odds model is that it potentially results in a large number of coefficients to be estimated, if either the number of outcome categories or the number of predictors is large. The total number of coefficients (including intercepts) to be estimated in a non-proportional odds model is $(J - 1) \times (k + 1)$, where J is the number of outcome categories and k the number of predictors. For example, in a model predicting an outcome with five categories via 10 predictors, we would need to estimate 44 coefficients (four intercepts and 40 slopes). This may make the model difficult to interpret and may even make it difficult to estimate when the sample is not sufficiently large. In comparison, a proportional odds model for an outcome with five categories and 10 predictors would only feature 14 coefficients (four intercepts and 10 slopes).

What is more, although the proportional odds assumption may not hold for all predictors, it may well hold for some. For those predictors where the proportional odds assumption holds, there is in fact only one population odds ratio, but a non-proportional odds model would yield several estimates, which is unsatisfactory. These considerations motivate the partial proportional odds model.

The partial proportional odds model

The partial proportional odds model is a special case of the general ordinal logit model, in which we assume proportional odds for some, but not all, predictors.

In other words, in a partial proportional odds model, we constrain the coefficients of some of our predictors to be the same across all dichotomisations of the outcome (as in the proportional odds model), but for other predictors we do not (as in the non-proportional odds model).

For example, given the result of Brant's test for Model 3.5, we may wish to fit a model that assumes proportional odds for gender, age and age squared, but not for *LOGFRDS*. Such a partial proportional odds model looks as follows:

Model 3.7: Partial proportional odds model

$$\text{logit}\left[P\left(LIFESAT_i > 'low'\right)\right] = \alpha_1 + \beta_1 Male_i + \beta_2 AgeC_i + \beta_3 AgeC_i^2 + \beta_{41} LOGFRDS$$

$$\text{logit}\left[P\left(LIFESAT_i > 'medium'\right)\right] = \alpha_2 + \beta_1 Male_i + \beta_2 AgeC_i + \beta_3 AgeC_i^2 + \beta_{42} LOGFRDS$$

Notice that *Male, AgeC* and *AgeC²* have the same slope coefficients (β_1, β_2, and β_3) in both equations, reflecting the assumption of proportional odds for these predictors. *LOGFRDS*, however, has a different coefficient for each equation (β_{41} and β_{42}), so we don't assume proportional odds for this predictor.

Table 3.11 shows coefficient estimates from the proportional odds, non-proportional odds and partial proportional odds models (Models 3.5, 3.6 and 3.7). We see that the partial proportional odds model estimates seven coefficients: two intercepts; one slope coefficient each for *AgeC, AgeC²* and *Male*; and two slope coefficients for *LOGFRDS*. The two coefficient estimates for *LOGFRDS*, although both positive, are appreciably different from each other, corresponding to odds ratios of 1.85 and 1.32, respectively. This suggests that a 1-unit difference in *LOGFRDS* is estimated to increase the odds of having higher than 'low' life satisfaction by a factor of 1.85, while the same difference increases the odds of having higher than 'medium' life satisfaction by a factor of 1.32 (controlling for age and gender). So the strength of the association between *LOGFRDS* and life satisfaction depends on which levels of life satisfaction we compare. More generally, it is possible that in a partial proportional odds model, coefficients for variables may be very much different, and even have opposite signs. Williams (2016) gives a nice example.

Table 3.11 Coefficient estimates (and standard errors) for three ordinal logistic regression models predicting life satisfaction by age, age squared, gender and number of friends

	Model 3.5: Proportional Odds	Model 3.7: Partial Proportional Odds		Model 3.6: Non-Proportional Odds	
	LIFESAT > j	LIFESAT > Low	LIFESAT > Medium	LIFESAT > Low	LIFESAT > Medium
Male	−0.250 (0.082)	−0.245 (0.082)	—	−0.313 (0.127)	−0.224 (0.087)
AgeC	0.109 (0.025)	0.111 (0.025)	—	0.097 (0.039)	0.111 (0.026)
AgeC²	0.041 (0.013)	0.040 (0.013)	—	0.021 (0.021)	0.045 (0.014)
LOGFRDS	0.363 (0.070)	0.618 (0.103)	0.278 (0.075)	0.615 (0.103)	0.279 (0.075)
Intercepts					
α_1	1.306 (0139)	0.907 (0.177)		0.992 (0.188)	
α_2	−0.788 (0.136)		−0.640 (0.143)		−0.664 (0.144)

Note: Standard errors for coefficients are shown in brackets. In the partial proportional odds model, the coefficients for *Male, AgeC* and *AgeC²* for the dichotomisation *LIFESAT* > medium are by definition identical to their counterparts in the dichotomisation *LIFESAT* > low. *AgeC*: Continuous age centred on 50 and measured in decades. *AgeC²*: continuous age squared. *LOGFRDS* = log(1 + friends). *LIFESAT* = life satisfaction. Data were taken from the Faith Matters Survey, 2006.

Looking now at the non-proportional odds model, we see that in this model all predictors have two different estimated coefficients. If the proportional odds assumption is in fact true for a variable, the two coefficients for that variable in the non-proportional odds model are the same in the population, but are of course not necessarily exactly identical in a particular data set. For *Male, AgeC* and *AgeC²*, the differences between the two estimates are small, consistent with the idea that for these three predictors the proportional odds assumption may well hold.

I would like to draw your attention to another aspect of Table 3.11, namely, the standard errors (given in brackets) and how they vary between the three models. Take *Male* as an example: the two standard errors for *Male* in the non-proportional odds model each are larger than the standard error for the single coefficient for *Male* in the two other models. Usually, whenever we relax the proportional odds assumption for a predictor, we obtain larger estimated standard errors for that predictor compared to a model where we assume proportional odds. Larger standard errors mean less precision in our estimates, and thus wider confidence intervals and possibly lower statistical power for hypothesis tests. So, by making the proportional odds assumption, we buy ourselves greater precision for our estimates. Conversely, the price for freedom from the proportional odds assumption is a loss in precision.

Likelihood ratio test comparing proportional odds, partial proportional odds and non-proportional odds models

The three models 3.5, 3.6 and 3.7 are nested within one another. The proportional odds model is nested within the partial proportional odds model, because we can turn the latter into the former by adding the restriction $\beta_{41} = \beta_{42}$, that is, by setting the two coefficients for *LOGFRDS* to be the same. Analogously, we can turn the non-proportional odds model into the partial proportional model by setting the coefficients for *Male, AgeC* and *AgeC²* to be the same (i.e. we set $\beta_{11} = \beta_{12}$, $\beta_{21} = \beta_{22}$ and $\beta_{31} = \beta_{32}$). Because the three models are nested, we can perform likelihood ratio tests to formally compare them. The results of these likelihood ratio tests are displayed in Table 3.12. In each case, we test the null hypothesis that the smaller model fits the data as well as a larger one.

Table 3.12 Likelihood ratio test of Model 3.7 versus 3.5, and Model 3.6 versus 3.7

Model		Log Likelihood	Λ	Parameters	df	df Difference	p
3.5	Proportional odds	−2164.7		6	2221		
3.7	Partial proportional odds	−2160.0	9.3	7	2220	1	0.002
3.6	Non-proportional odds	−2158.9	2.2	10	2217	3	0.540

Note. Two likelihood ratio tests are displayed. The first compares the partial proportional odds model to the proportional odds model (second row). The other compares the non-proportional odds model to the partial proportional odds model (third row). *df* = degrees of freedom.

First, we compare the (full) proportional odds with the partial proportional model. Here, the likelihood ratio test statistic yields a very small p-value ($p = 0.002$), which suggests that the partial proportional odds model explains the data better than the proportional odds model. The second likelihood ratio test compares the partial proportional odds model with the non-proportional odds model. This second test yields a large p-value ($p = 0.540$), suggesting that there is little indication that the non-proportional odds model provides a better explanation of life satisfaction than the partial proportional odds model. Overall, then, these findings suggest that, out of the three models investigated, the partial proportional odds model should be preferred.

Conducting likelihood ratio tests in this manner is a principled way of comparing non-proportional, partial proportional and full proportional odds models. A test of the non-proportional versus the proportional odds model is essentially testing the same null hypothesis as an omnibus Brant's test conducted on the coefficients from the proportional odds model and should in general yield similar results. The usefulness of the Brant's test is in identifying the covariates for which the proportional odds assumption appears to be violated and for which it appears to hold.

Ordinal logistic regression in action

This section walks through a complete example of an ordinal logistic regression, considering particularly the choice between a 'pure' proportional odds model and a partial proportional odds model. We will once again use data from the Faith Matters Survey. In their analysis, Lim and Putnam (2010) started from the assumption, based on previous research in the field, that religiosity is associated with life satisfaction and that this relationship cannot be completely explained by other variables that both religiosity and life satisfaction are related to. They wished to know what explains this association: does religious identity or religious feeling itself lead to higher life satisfaction, or could it be the belonging to a religious community that matters? This chapter will present a similar analysis to Lim and Putnam's, but the example is simplified a bit to make it easier to explain. Our outcome is still life satisfaction, but this time I have coded this variable into six ordinal categories (from 'extremely dissatisfied' to 'extremely satisfied'), rather than three categories as in our earlier examples. Box 3.2 gives a summary of the variables used in the models that follow.

Let's suppose our main research questions are whether life satisfaction is related to religious identity and to the number of friends a person has in their religious congregation (provided they belong to one). The number of congregational friends is intended as a measure of a person's sense of belonging to a religious community, which represents the communal aspect of religion, while religious identity is taken here to represent the individual aspect.

To investigate these questions, we might propose the following model:

Model 3.8: No interaction, proportional odds

$$\text{logit}\left[P(LIFESAT_i > j)\right] = \alpha_j + \beta_1 RELID_i + \beta_2 LOG.CNGFRDS_i + \text{covariates},$$
$$j = \{1, 2, 3, 4, 5\}$$

where *RELID* is a dummy variable identifying people with a strong religious identity (as defined previously in this chapter) and *LOG.CNGRDS* is the number of friends respondents report to have in their religious congregation (0 by definition for those that don't attend religious services), which was log transformed in the same way as *LOGFRDS* (see Box 3.2, and also Table 3.9 above). Covariates are other predictors that we wish to control for. In our example, the covariates we consider are gender and age. All variables are described in Box 3.2. Because we use the six-category life satisfaction variable as our outcome, there are five cutpoints *j*.

Box 3.2

Variables used in Models 3.8(a) and 3.9(a)

Life Satisfaction: Life satisfaction was measured on a 10-point ordinal response scale, from 1 = *extremely dissatisfied* to 10 = *extremely satisfied*. More than 85% of respondents rated themselves as either 7, 8, 9, or 10. Few people used categories 1 to 6, which led to problems with sparseness when trying to estimate ordinal logistic models (see Chapter 2, section 'Things That Might Go Wrong', for a discussion of sparseness). I have therefore combined the lower scale points, so that ratings 1 to 4 form the lowest category and ratings 5 to 6 the next lowest. Ratings 7 to 10 are kept as separate categories. Thus, we end up with a six-category classification of life satisfaction. There is nothing particularly complicated about using an outcome with many categories. It simply means that we have a large number of cutpoints, and hence a large number of intercepts.

CNGFRDS and *FRIENDS* both were measured using the response scale: 'none', '1–2', '3–5', '6–10' and '11 or more'. Both variables were transformed into a numeric variable by taking the midpoint of the categories given: 'none' = 0, '1–2' = 1.5, '3–5' = 4, '6–10' = 8 and '11 or more' = 11. It also turned out (not shown here) that log transforming these variables improved the fit of the model, so the variables $LOGFRDS = \log(FRIENDS + 1)$ and $LOG.CNGFRDS = \log(CNGFRDS + 1)$ were constructed and used as predictors.

(Continued)

RELID is a dummy variable measuring religious identity, coded 1 for respondents who said that religion was 'very important' in their lives, and 0 for others.

Age was centred on its median (50 years) and the result divided by 10 in order to make the coefficients more interpretable. The odds ratio for the resulting variable (*AgeC*) represents the estimated effect on the odds of being in a higher life satisfaction category for a 10-year difference in age. As in previous analyses in this chapter, we include both a linear and a square term (*AgeC²*) for age in the model.

Male is a dummy variable coded 1 for males and 0 for females.

Many more covariates could and should be considered, such as religious denomination and attendance of religious services. We keep the example simple so that we can focus on the technical statistical aspects. If you are interested in the full model used by Lim and Putnam, have a look at their 2010 publication.

In addition, we may be interested in whether religiosity matters to the same extent regardless of the communal aspect of religion. Does religiosity give the same benefit in terms of life satisfaction at every level of religious belonging? We might have the hypothesis that religious identity has a stronger benefit among those who are strongly integrated into their congregation (as represented by the number of congregational friends) than among those who are not. To investigate this hypothesis, we propose a model that allows an interaction between *RELID* and *CNGFRDS*:

Model 3.9: With interaction, proportional odds

$$\text{logit}\left[P(LIFESAT_i > j)\right] = \alpha_j + \beta_1 RELID_i + \beta_2 LOG.CNGFRDS_i +$$
$$\beta_3 \left(RELID_i \times LOG.CNGFRDS_i\right) + \text{covariates}$$

To gauge the statistical evidence for the presence of an interaction between *RELID* and *LOG.CNGFRDS*, we could estimate both Models 3.8 and 3.9 and compare the two, using a likelihood ratio test. However, we might be concerned that the proportional odds assumption may not hold for some of the predictors in the model. Therefore, we might first estimate the larger model (Model 3.9) and conduct Brant's tests to investigate the plausibility of the proportional odds assumption. Table 3.13 shows the results of these Brant's tests.

Brant's omnibus test returns a small p-value ($p = 0.016$), indicating that the proportional odds assumption may not hold for at least one predictor. The individual Brant's test results suggest that the problem lies with *LOGFRDS*, whose individual Brant's test yields $p < 0.001$. All other individual Brant's tests yield large p-values, suggesting that there is little indication of a violation of the proportional odds assumption for variables other than *LOGFRDS*. We thus choose to relax the proportional odds assumption

Table 3.13 Brant's tests of the proportional odds hypothesis for Model 3.9

	Test Statistic (Chi-Square)	df	p
Omnibus	46.4	28	0.016
Male	2.3	4	0.687
AgeC	1.4	4	0.847
AgeC²	4.5	4	0.344
LOG. CNGFRDS	5.7	4	0.223
RELID	5.8	4	0.216
RELID × LOG. CNGFRDS	1.9	4	0.752
LOGFRDS	20.1	4	0.000

Note. Age.C = continuous age; Age.C² = continuous age squared; df = degrees of freedom; LOG. CNGFRDS = log congregational friends; RELID = religious identity; LOGFRDS = log transformed; RELID = religious identity.

for *LOGFRDS* and estimate partial proportional odds models to address our research question. Our two partial proportional odds models are as follows:

Model 3.8a: No interaction, partial proportional odds

$$\text{logit}\left[P(LIFESAT_i > j)\right] = \alpha_j + \beta_1 RELID_i + \beta_2 LOG.CNGFRDS_i + \beta_{3j}LOGFRDS_i + \text{covariates}$$

Model 3.9a: With interaction, partial proportional odds

$$\text{logit}\left[P(LIFESAT_i > j)\right] = \alpha_j + \beta_1 RELID_i + \beta_2 LOG.CNGFRDS_i + \beta_{3j}LOGFRDS_i + \beta_4\left(RELID_i \times LOG.CNGFRDS_i\right) + \text{covariates}$$

In both models, the coefficient for *LOGFRDS*, β_{3j}, has the additional subscript *j*, to indicate that the effect of *LOGFRDS* is allowed to be different for each cutpoint.

The results of a likelihood ratio test comparing these two models are displayed in Table 3.14.

Table 3.14 Likelihood ratio test of Model 3.9a versus Model 3.8a

Model		LL	Λ	Parameters	df	df Difference	p
3.8a	No interaction	−3513.8		15	2212		
3.9a	Interaction: *RELID × LOG.CNGFRDS*	−3511.9	3.92	16	2211	1	0.048

Note. RELID = religious identity; LOG.CNGFRDS = log congregational friends; LL = log likelihood; df = degrees of freedom.

The likelihood ratio test yields a *p*-value of $p = 0.048$. This suggests that there is some evidence that the larger model (Model 3.9a) might improve the prediction of life satisfaction relative to the smaller model (Model 3.8a). In this example, this means that there is some statistical evidence for an interaction between *RELID* and *LOG.CNGFRDS* in the population. Let's therefore investigate the results of Model 3.9a. The estimated coefficients and odds ratios are displayed in Table 3.15.

Table 3.15 Estimates from a partial proportional odds model predicting life satisfaction (Model 3.9a)

Predictors	Estimate	SE	OR	95% CI for OR
Male	−0.183	0.077	0.83	[0.71, 0.97]
AgeC (in decades)	0.076	0.024	1.08	[1.03, 1.13]
AgeC²	0.033	0.012	1.03	[1.01, 1.06]
LOG.CNGFRDS	0.239	0.071	1.27	[1.10, 1.46]
RELID: very important	0.235	0.137	1.26	[0.97, 1.65]
RELID × *LOG.CNGFRDS*	0.182	0.091	1.20	[1.002, 1.44]
LOGFRDS				
LIFESAT > 1	0.788	0.173	2.20	[1.57, 3.04]
LIFESAT > 2	0.476	0.106	1.61	[1.29, 2.00]
LIFESAT > 3	0.311	0.086	1.36	[1.14, 1.63]
LIFESAT > 4	0.130	0.078	1.14	[0.97, 1.33]
LIFESAT > 5	0.030	0.085	1.03	[0.87, 1.22]
Intercepts				
LIFESAT > 1 (α_1)	1.523	0.279		
LIFESAT > 2 (α_2)	0.589	0.196		
LIFESAT > 3 (α_3)	−0.045	0.171		
LIFESAT > 4 (α_4)	−1.006	0.165		
LIFESAT > 5 (α_5)	−1.623	0.177		

Note. AgeC: The age variable was centred at the median (50 years) and divided by 10, so that the coefficient represents the effect associated with a 10-year age difference. *AgeC²*: The square of the age variable as defined above. *LOG.CNGFRDS* = log congregational friends; *RELID* = religious identity; *LOGFRDS* = log friends; Life Satisfaction (*LIFESAT*) was coded into six categories, numbered from 1 to 6. The highest category is 6: 'extremely satisfied'. 95% confidence intervals were calculated using the profile likelihood method. *SE* = standard error; *OR* = odds ratio; *CI* = confidence interval. Data were taken from the Faith Matters Survey.

Before we interpret the results in the light of the research questions, I wish to make a remark about the formal aspects of Table 3.15. Since our outcome, life satisfaction, has six categories, we now have five cutpoints, and thus have five intercepts in the bottom part of the table. Because we allowed non-proportional odds for the predictor *LOGFRDS*, we have five coefficient estimates associated with this predictor.

The variable *LOGFRDS* is not of primary interest in our investigation in this example, but in order to show how we might interpret estimates of non-proportional odds, let's have a look at the estimated odds ratios for *LOGFRDS*. Consider first the odds ratio for '*LIFESAT* > 1' – which estimates the effect of *LOGFRDS* on the odds of having better than 'extremely dissatisfied' life satisfaction. This odds ratio is estimated to be 2.20, suggesting a relatively large effect. This suggests that having a larger number of friends is associated with a smaller probability of having very low levels of life satisfaction. Looking at the remaining four odds ratios for *LOGFRDS*, we see that they get closer to 1 as we move up the cutpoints. In fact, the last two odds ratios are statistically indistinguishable from 1, judging by their 95% CIs. This suggests that the number of friends is relatively less important for predicting the difference between moderately high and high life satisfaction and is essentially unrelated to the difference between the top three life satisfaction categories. To interpret the overall pattern in less formal, and slightly imprecise language, these results suggest that having more friends may protect you against misery, but it won't necessarily lift you from moderate satisfaction to deep fulfilment.[viii]

Let's now turn to the main purpose of Lim and Putnam's (2010) investigation, which concerns the relationships between life satisfaction and the number of *congregational* friends, as well as between life satisfaction and religious identity. Our interpretation of the effects of these variables needs to take account of the interaction between them, which we have included in the model. First, let's carefully go through the numbers and their meaning:

- The odds ratio for *LOG.CNGFRDS* tells us the estimated effect associated with a 1-unit change in this variable *among those for whom religion is not very important* (who have *RELID* = 0). This odds ratio is estimated to be 1.26, with a 95% CI from about 1.10 to 1.46.
- The odds ratio for RELID tells us the difference between those who find religion very important versus those who do not – *among those who report having no congregational friends*. This odds ratio is estimated to be 1.27, although the 95% CI [0.97, 1.65] includes the value 1, suggesting that from these data we haven't got strong evidence for an effect of *RELID* among those with no congregational friends.

[viii]This formulation is of course imprecise and not entirely justified by the data. Whether being 'extremely satisfied' is the same thing as 'deep fulfilment' is a question too profound for a statistics textbook to discuss competently. Besides, of course, this analysis on its own does not prove the direction of the causal association between life satisfaction and the number of friends. My formulation assumes that having friends contributes to higher life satisfaction, but the causal direction may well (also) work the other way around. For example, extremely dissatisfied people may tend to lose their friends, or may tend not to make friends in the first place. It is also plausible that other factors, such as mental illness or being unemployed, are related both to the number of friends and to life satisfaction, and that such factors play a part in causing the association we observe.

- The interaction *RELID* × *LOG.CNGFRDS* is estimated to be 1.20, which suggests that among those for whom religious identity is very important, the effect of *LOG. CNGFRDS* is substantially higher than among those for whom it is not. In fact, the odds ratio of *LOG.CNGFRDS* for those who think religion is important to them is estimated to be 1.26 × 1.20 ≈ 1.5.

These results suggest that religious identity might not necessarily be associated with higher life satisfaction in the absence of congregational friends. However, when a religious identity is combined with the experience of a faith community, then religious identity does have an association with higher life satisfaction.

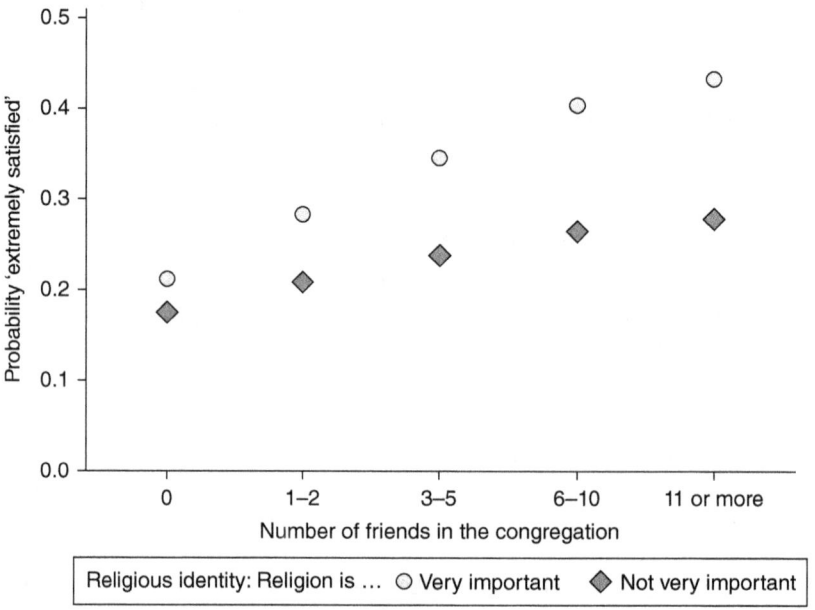

Figure 3.4   The predicted probability of being 'extremely satisfied' as a function of religious identity and number of congregational friends, according to Model 3.9a

Notes. The figure displays the estimated probabilities to be 'extremely satisfied' for a female respondent aged 50 years who reports having 11 or more 'total' friends (inside or outside of any religious congregation).

To illustrate this result further, I have visualised the estimated effects of religious identity, congregational friends and their interaction in Figure 3.4. To do this, I calculated the predicted probabilities of being in the highest life satisfaction category (*LIFESAT* = 6), for a female respondent of median age (50 years) who reported having 11 or more friends, using all possible combinations of the variables 'congregational friends' and 'religious identity'.[ix] We see that the number of congregational friends makes relatively little difference to the probability of being 'extremely satisfied' for those without strong religious identity, but makes a bigger difference for those with a strong religious identity.

In the research publication that inspired this example, Lim and Putnam (2010) report essentially the same result and illustrate it with a graph similar to Figure 3.4. However, they control for an even greater number of covariates (e.g. attendance at religious services and religious denomination) and use appropriate sampling weights in their analysis. They find that the number of congregational friends has essentially no effect at all for those without strong religious identity. They frame their conclusion thus: 'These findings suggest that in terms of life satisfaction, it is neither faith nor communities, per se, that are important, but communities of faith. For life satisfaction, praying together seems to be better than either bowling together or praying alone' (p. 927).

From the results of Model 3.9a alone, this conclusion is not justified. It is in principle entirely plausible to argue that the relationship between the number of congregational friends and life satisfaction may work the other way around: that it is due to a tendency of more satisfied people to make more congregational friends. But Lim and Putnam (2010) do not justify their conclusion with just a single statistical model. They also review other studies conducted on the topic and conduct further analyses which aim to establish the direction of causality. I won't discuss those analyses here, but wish to emphasise the general point that the interpretation of the results of estimating a statistical model on a data set should take many things into account, including research design, methodology, theoretical considerations, and knowledge that we have from other research.

Chapter Summary

In the social sciences, we frequently encounter ordinal variables. In particular, many survey questions that inquire after attitudes, opinions or values have ordinal response scales (e.g. from 'strongly agree' to 'strongly disagree'). It may thus be surprising that ordinal logistic regression is not more frequently used. One reason for this is that survey questions with ordinal response scales are often combined to form Likert scales, where scores on individual items are summed, thus implicitly treating them as if they were interval-level variables.

(Continued)

[ix]I could also have chosen other values for the covariates, such as '30 years old', 'male' and 'has no friends'. This would have changed the predicted probabilities, of course, but would have made little difference to the relative differences between the groups defined by religious identity and the number of congregational friends.

Lack of familiarity with ordinal regression may be another reason why this method is not used more often by social scientists. At the time that I am writing this book, many researchers understand the term *ordinal logistic regression* to refer to the proportional odds model only and are not aware of the generalised ordered logit model that encompasses the three model types that we have discussed in this chapter: proportional odds, partial proportional odds, and non-proportional odds. For example, Lim and Putnam (2010) mention that the proportional odds assumption is violated for their variable 'number of friends' and discuss this as a potential limitation of their analysis; but they don't consider a partial proportional odds model, which would have alleviated the limitation. Publications by Williams (2006, 2016) have helped to make the generalised ordered logit model and the partial proportional odds model better known. Williams (2016) in addition criticises that some researchers, even when they do use the partial proportional odds model, don't consider how the non-proportional odds in some of their predictors might be interpreted. In the section 'Ordinal Logistic Regression in Action', I gave an example of how we might understand the non-proportional odds in the relationship between life satisfaction and the number of friends.

Generalised ordinal logit regression is not the only type of model for ordinal outcomes. An alternative is ordinal probit regression, which is in many ways similar to ordinal logistic regression, but assumes that the ordinal categories are imprecise measurements of a latent variable that is normally distributed. See, for example, Agresti (2013) for more detail on this.

Further Reading

Agresti, A. (2013). *Categorical data analysis* (3rd ed.). Wiley.

Hilbe, J. M. (2009). *Logistic regression models*. Chapman & Hall.

If you want to read a mathematically and technically more elaborate description of ordinal logistic regression, consider the above texts.

Long, J. S., & Freese, J. (2014). *Regression models for categorical dependent variables using Stata* (3rd ed.). Stata Press.

This book offers you detailed guidance on how to conduct an ordinal logistic regression in Stata.

Williams, R. (2016). Understanding and interpreting generalized ordered logit models. *Journal of Mathematical Sociology, 40*(1), 7–20.

A particularly good explanation of the partial proportional odds model can be found in this article.

4

MULTINOMIAL LOGISTIC REGRESSION

Chapter Overview

This chapter turns to the third member of the logistic regression family of models covered in this book. Multinomial logistic regression is a model for outcome variables that are nominal and have more than two categories. In contrast to ordinal variables, nominal variables have categories that cannot be placed in a meaningful order. Some examples of nominal variables that social scientists might be interested in modelling are:

- Choice of undergraduate study subject, with categories such as 'science', 'humanities', and 'arts'
- Pupils' plans after leaving school: 'enrol in further study', 'get a paid job', 'start an apprenticeship', 'other plans', and 'no plan yet'
- Type of accommodation: 'rented', 'owned with mortgage', 'owned outright', 'nursing home or other institution', 'homeless', and 'other'

The following conditions must be met by a nominal outcome variable for multinomial logistic regression to be appropriate:

- The outcome categories must be mutually exclusive. Taking the example of school leavers' plans, we assume that each pupil's plan can be classified into exactly one of the five available categories.[i]
- The number of outcome categories should be small relative to the number of cases in the data set. In outcomes with too many categories, it is likely that some categories occur rarely. Rare outcome categories lead to sparseness in the data and are associated with the same estimation problems that we discussed in Chapter 2 in the context of binary logistic regression (see the section 'Things That Might Go Wrong: Estimation Problems' in Chapter 2).

Example data for multinomial logistic regression

To illustrate the analysis of a nominal outcome variable, we return to the Family Survey Dutch Population (FSDP) data from 1998 (de Graaf et al., n.d), which we already analysed in Chapter 2. Kraaykamp et al. (2010) were interested in predictors of cultural consumption in the Netherlands. Their aim was to test sociological theories about the social display of status, which predict that people with high social status would tend to signal this status by engaging in 'highbrow' cultural activities, such

---

[i]In practice, we may encounter cases where categories of nominal variables are not mutually exclusive. For example, a school leaver may plan to combine paid work with enrolling in further study. How the researcher deals with this issue will depend on the aim of the investigation. Some options in this example are: require respondents to identify only one 'main' plan, define combinations as categories in their own right ('study + work'), or analyse each category separately as a binary variable ('work' vs. 'not work'; 'study' vs. 'not study' etc.), using binary logistic regression.

as going to the theatre, classical music concerts, and museums,[ii] as well as by avoiding 'lowbrow' cultural activities, such as pop concerts. We will consider these theories and how to test them later in this chapter, in the section 'Multinomial Logistic Regression in Action'. For now, let us consider the outcome variable of Kraaykamp et al.'s investigations. Based on their respondents' answers to a series of survey questions about their cultural activities, Kraaykamp et al. (2010) classified each respondent into one of four types of cultural consumer:

1   Highbrow: attends at least one highbrow activity (theatre, museums and/or classical concerts) but does not attend pop concerts
2   Omnivore: attends at least one highbrow activity, and also attends pop concerts
3   Pop only: attends pop concerts only (Kraaykamp et al., 2010, call this category 'univores')
4   Inactive: does not attend any of the four types of events[iii]

This scheme is summarised in Table 4.1.

Table 4.1   How the cultural consumption variable was constructed

Cultural Consumption Category	Attends One or More of • **Museums** • **Theatre** • **Classical Concerts**	Attends Pop Concerts
Omnivore	Yes	Yes
Highbrow	Yes	No
Pop only	No	Yes
Inactive	No	No

Note. 'Attendance' was defined as respondents reporting that they attended a given type of event once per year, or more often.

Before we investigate the relationship between social status and consumption, let's start with a simple example, in order to get to know our outcome variable. Table 4.2 gives an overview of the distribution of the cultural consumption variable in a subset of the FSDP data from 1998.[iv] It also shows how cultural consumption varies by area of residence of the respondent (urban or rural).

[ii]There are obviously different types of museums, not all of which may be considered unambiguously 'highbrow'. However, Kraaykamp et al. (2010) argue that at the time their data were collected, most museums in the Netherlands were seen as part of 'highbrow' culture.

[iii]Kraaykamp et al.'s (2010) cultural consumption variable was constructed from responses to several survey questions. But of course other nominal variables may well result from responses to just one survey question.

[iv]Kraaykamp et al. (2010) use data from several surveys conducted in the Netherlands between 1992 and 2003, and also use information from both the primary survey respondents and their married or cohabiting partners. I present here a simplified analysis using the 1998 survey and primary respondents only, to make the example easier to explain.

Table 4.2 Cultural consumption patterns and rural versus urban residence in the Family Survey Dutch Population, 1998: counts and percentages

	Residence					
Cultural Consumption	**Urban**		**Rural**		**Total**	
Omnivore	234	37.6%	25	26.6%	259	36.1%
Highbrow	279	44.8%	44	46.8%	323	45.0%
Pop only	30	4.8%	8	8.5%	38	5.3%
Inactive	80	12.8%	17	18.1%	97	13.5%
Total	623	100.0%	94	100.0%	717	100.0%

Note. Data were taken from the Family Survey Dutch Population 1998 (de Graaf et al., n.d.).

In both urban and rural areas, the 'highbrows' are the biggest group of cultural consumers, followed by the omnivores and the inactives; the 'pop only' respondents are the smallest group. Nonetheless, urban and rural areas differ in the proportions of respondents in each outcome category: the proportion of omnivores is higher in urban than in rural areas, while the proportions of highbrows, pop only respondents and inactives is higher in rural than in urban areas. Another way of saying the same thing is that being an omnivore is *relatively more probable* in urban compared to rural areas, while being inactive, or being a highbrow or pop only cultural consumer, is relatively more probable in rural areas, compared to urban areas. The next section explores this notion of relative probability, or relative risk, in more detail.

Relative risks

Categories of multinomial outcomes can be compared using **relative risks**. 'Risk' here is just another name for probability. A relative risk (*RR*) is the probability of being in one outcome category divided by the probability of being in another.[v] For example, let's consider the probability of being highbrow compared to the probability of being an omnivore. Taking the total sample from Table 4.2, the observed relative risk ($\widehat{RR}$) is:

$$\widehat{RR}_{highbrow/omnivore} = \frac{\hat{p}_{highbrow}}{\hat{p}_{omnivore}} = \frac{323/717}{259/717} = 1.25$$

---

[v]In Chapter 2, section 'Risk Ratio and Absolute Risk Difference', we encountered the related concept of the 'risk ratio', a statistic that summarises the findings of a 2 × 2 table. The risk ratio in a 2 × 2 table is essentially a special case of a relative risk, in a situation where the outcome has exactly two categories. Note also that the term 'relative risk' has a different meaning in epidemiology, where it refers to the ratio of the risks of disease between two groups.

where $\hat{p}_{highbrow}$ is the estimated probability of being a highbrow cultural consumer. We estimate this probability by the observed proportion of highbrows in our data set – that is $\hat{p}_{highbrow} = \frac{323}{717} = 0.45$. Analogously, $\hat{p}_{omnivore}$ is the estimated probability of being an omnivore. As usual for sample statistics, we put a 'hat' on p and RR, to indicate that these are estimates from a particular data set, not the true probabilities and not the true relative risk in the population.

The result of our calculation means that the observed proportion of being highbrow in this data set is 1.25 times higher than the observed proportion of being an omnivore. Equivalently, we could say that the probability of being highbrow is estimated to be 25% higher than the estimated probability of being an omnivore.

The calculation of relative risks looks similar to the calculation of odds, but there is an important difference. In the context of multinomial regression, a relative risk is the probability of an event divided by the probability of another event, when yet other events are also possible outcomes. In contrast, odds are the probability of an event divided by the probability of the same event not occurring. For example, the observed odds of being 'highbrow' in Table 4.2 are:

$$\widehat{Odds}_{highbrow} = \frac{\hat{p}_{highbrow}}{1 - \hat{p}_{highbrow}} = \frac{323/717}{1 - 323/717} = \frac{0.45}{1 - 0.45} = 0.82$$

So the odds of being 'highbrow' are quite a different thing than the relative risk of being highbrow versus being an omnivore.

Turning our attention back to relative risks, we may also be interested in describing how a relative risk depends on a predictor variable. We can estimate a conditional relative risk, given a fixed value of a predictor variable. To do this, we use conditional probabilities in the calculation. (On the concept of conditional probability, see Chapter 2, section 'Probabilities and Conditional Probabilities'.) For example, the relative risk of being an omnivore compared to being inactive among urban respondents is as follows:

$$\widehat{RR}_{highbrow/omnivore} \mid Residence : urban = \frac{\hat{p}_{highbrow} \mid urban}{\hat{p}_{omnivore} \mid urban} = \frac{279/623}{234/623} = 1.19$$

The corresponding relative risk among rural respondents is:

$$\widehat{RR}_{highbrow/omnivore} \mid Residence : rural = \frac{\hat{p}_{highbrow} \mid rural}{\hat{p}_{omnivore} \mid rural} = \frac{44/94}{25/94} = 1.76$$

So we see that in this data set, the relative risk, highbrow versus omnivore, is higher among rural respondents (1.76) than among urban respondents (1.19). This suggests that rural residents are relatively more likely to choose being highbrow rather than being omnivorous, compared to urban residents. Next we introduce a formal measure of the relationship between a predictor and a relative risk.

Relative risk ratios

We need a measure of the strength of the association between a predictor and the relative risk of two nominal outcome categories. Such a measure is the **relative risk ratio** (*RRR*). As the name implies, this is the ratio of two relative risks. For example, the observed relative risk ratio of rural versus urban residents, for being highbrows versus omnivores, is as follows:

$$\widehat{RRR}^{high}\!\big/_{omni},\ ^{rur}\!\big/_{urb} = \frac{\widehat{RR}^{high}\!\big/_{omni}\ |\ rural}{\widehat{RR}^{high}\!\big/_{omni}\ |\ urban}$$

$$= \frac{1.76}{1.19}$$

$$= 1.48$$

We might interpret this as follows: the relative risk of being a highbrow cultural consumer rather than an omnivore is 1.48 times higher (or 48% higher) among the rural respondents than among the urban respondents in this sample.

Generally, when we analyse how the relative risk of outcome category A versus outcome category B depends on a predictor variable, then

- *RRR* > 1 indicates that the predictor is associated with a higher relative risk of category A versus category B,
- *RRR* < 1 indicates that the predictor is associated with a lower relative risk of category A versus category B, and
- *RRR* = 1 indicates that there is no relationship between a predictor and the relative risk of the two outcome categories.

Above, we have calculated the relative risk ratio of highbrows versus omnivores, for rural versus urban respondents. In the same way, we could in principle consider all possible pairs of outcome categories (omnivore vs pop only, highbrow vs inactive, and so forth) and calculate the relative risks, as well as the relative risk ratios of urban versus rural residents. However, this is not usually done in practice. For the purpose of this chapter, the importance of relative risks and relative risk ratios derives from their application within the multinomial regression model. The next section gives a first example of this model.

A simple example of multinomial logistic regression

Let's now see how relative risks and relative risk ratios are modelled in multinomial logistic regression. The multinomial logistic regression model is set up in a manner very similar to binary logistic regression (see Chapter 2). The difference is that in binary logistic regression we model the *log odds* of an event, whereas in multinomial logistic regression we model the *log relative risk* of one event versus another (i.e., of one outcome category versus another).

Let's start with a simple example and propose a multinomial regression model predicting cultural consumption via just one predictor, the binary variable 'residence' (urban vs rural). We begin by choosing a reference category from our four outcome categories. Let's choose 'omnivores' for now. (I say more about how to choose a reference category the section 'Multinomial Regression: Some Additional Comments'.) Our model aims to predict the log relative risks of being in each of the other categories compared to the reference category. The model consists of three equations and looks like this:

Model 4.1:

$$\log\left(\widehat{RR}^{high}\!\big/_{omni}\right) = \alpha_1 + \beta_1 Rural_i$$

$$\log\left(\widehat{RR}^{pop}\!\big/_{omni}\right) = \alpha_2 + \beta_2 Rural_i$$

$$\log\left(\widehat{RR}^{inact}\!\big/_{omni}\right) = \alpha_3 + \beta_3 Rural_i$$

where *Rural* is a dummy variable coded 0 = urban and 1 = rural.

Each equation predicts the logarithm of a relative risk of one outcome category versus the reference category. Each of the three equations has its own intercept (α_1, α_2, α_3), and its own slope coefficient for *Rural* (β_1, β_2, β_3).

Table 4.3 shows the results from estimating Model 4.1 on the FSDP data shown in Table 4.2.

Table 4.3   Results from a multinomial logistic regression of cultural consumption on area of residence

Category (vs Omnivore)		Estimate	SE	RRR	95% CI for RRR
Highbrow	Intercept (α_1)	0.176	0.089		
	Rural (β_1)	0.389	0.266	1.48	[0.88, 2.51]
Pop only	Intercept (α_2)	−2.054	0.194		
	Rural (β_2)	0.915	0.450	2.50	[0.98, 5.84]
Inactive	Intercept (α_3)	−1.073	0.130		
	Rural (β_3)	0.688	0.340	1.99	[1.01, 3.85]

Note. $N = 717$. 95% CIs were calculated by the profile likelihood method. *SE* = standard error; *RRR* = relative risk ratio; CI = confidence interval.

The results table shows the parameter estimates for the three model equations. For example, the equation for highbrow versus omnivore is as follows:

$$\log\left(\widehat{RR}_{high/omni}\right) = 0.176 + 0.389 \times Rural_i$$

This equation is presented on the scale of the log relative risk. To obtain a predicted relative risk, we can exponentiate both sides of the equation. The predicted relative risk for highbrow versus omnivore is as follows:

$$\widehat{RR}_{high/omni} = \exp(0.176 + 0.389 \times Rural_i)$$

We could write out the equations for the relative risks of univores versus omnivores and inactives versus omnivores in an analogous manner. If we wished, we could use these equations to calculate a predicted relative risk for each pair of outcome categories, separately for respondents living in urban and in rural areas. However, this is not often done in practice, because we can represent the model results more intuitively by calculating relative risk ratios.

The relative risk ratios for a predictor variable are calculated by exponentiating that predictor's slope coefficient. Thus, for example,

$$\widehat{RRR}_{high/omni}, Rural \; vs \; Urban = \exp(0.389) = 1.48$$

Thus, rural respondents are estimated to be about 1.48 times more likely to be highbrow cultural consumers, compared to omnivores, than urban respondents. However, the 95% CI for this relative risk ratio is [0.88, 2.51], which includes the value 1 and values smaller than 1, as well as values larger than 1. This indicates that the data are compatible with no difference, as well as a difference in either direction, between urban and rural populations in the relative risk to be highbrow compared to omnivore.

Note also that the model-estimated *RRR* is identical to the observed *RRR* that we calculated in the section 'Relative Risk Ratios'. This will always be the case for a model with a single categorical predictor, but it won't in general be true for models with a continuous predictor, or with more than one categorical predictor (unless all interactions between categorical predictors are included in the model).

We can also use the estimates from Model 4.1 to calculate predicted probabilities of being in each category. This is possible because, although the model parameters constitute relative probabilities, we know that the probabilities of each of the categories must sum to 1. I give a general formula for calculating predicted probabilities below, in the section 'Predicted Probabilities'.

Note that the choice of reference outcome category (here we chose 'omnivores') is arbitrary from a mathematical point of view. The predicted probabilities for each category will be the same regardless of the choice of reference category, if the model is otherwise the same. However, changing the reference category does change which relative risks are being explicitly compared. For example, if instead of 'omnivore' we made 'inactive' the reference category, we would estimate the relative risk ratios $RRR_{high/inact}$, $RRR_{omni/inact}$ and $RRR_{uni/inact}$. We can also get these RRRs from Table 4.3, although they are not directly displayed. For example,

$$RRR_{high/inact} = \frac{RRR_{high/omni}}{RRR_{inact/omni}} = \frac{1.48}{1.99} = 0.74$$

So rural respondents are estimated to be relatively less likely to be highbrow rather than inactive by a factor of 0.74, compared to urban respondents. We can also express the same thing by saying that rural respondents are relatively 26% less likely to be highbrow rather than inactive, compared to urban respondents.

The multinomial logistic regression model

Let's now consider the formal definition of a multinomial logistic regression model. The multinomial logistic regression model is:

$$\log\left(\frac{p_j}{p_J}\right) = \alpha_j + \beta_{j1}X_{1i} + \beta_{j2}X_{2i} + \ldots + \beta_{jk}X_{ki}, \quad j = 1,\ldots, J-1$$

where:

- The categories of the nominal outcome variable are numbered $j = 1, 2, \ldots J$, and the last (Jth) category is the reference category. Each of the categories $1, 2, \ldots,$ $J - 1$ is being compared to this reference in a separate model equation.
- p_j is the probability of being in the jth outcome category, and p_J is the probability of being in the reference category, J.
- $\frac{p_j}{p_J}$ is called the relative risk of category j compared to the reference category, J. We sometimes also write the relative risk as $RR_{j/J}$.
- α_j is the intercept of the jth equation; this is the log relative risk when all predictor variables are equal to zero.
- $\beta_{j1}, \beta_{j2}, \ldots, \beta_{jk}$ are the slope coefficients of the predictors $X_1, X_2, \ldots, X_k$, for the jth equation, where k is the number of predictor variables.

This way of writing the model is succinct, but it may not be very intuitive. The notation above really means that multinomial regression is a model that is made up of several equations. The number of equations is equal to the number of outcome categories minus 1. We call the number of outcome categories J, and so the number of equations is equal to $J - 1$. One of the outcome categories is chosen as the reference. For the purpose of mathematical notation, we make the last category the reference, so it gets the number J. Each model equation compares one of the other outcome categories to the reference category.

We could also write the multinomial regression model, with less succinct notation, as several equations:

$$\log\left(\frac{p_1}{p_J}\right) = \alpha_1 + \beta_{1,1}X_{1i} + \beta_{1,2}X_{2i} + \ldots + \beta_{1k}X_{ki}$$

$$\log\left(\frac{p_2}{p_J}\right) = \alpha_2 + \beta_{2,1}X_{1i} + \beta_{2,2}X_{2i} + \ldots + \beta_{2,k}X_{ki}$$

$$\ldots$$

$$\log\left(\frac{p_{J-1}}{p_J}\right) = \alpha_{J-1} + \beta_{J-1,1}X_{1i} + \beta_{J-1,2}X_{2i} + \ldots + \beta_{J-1,k}X_{ki}$$

Note that the predictor variables, X_1, X_2, . . . have a different slope coefficient in each equation. For example, the slope coefficients associated with X_1 are denoted by $\beta_{1,1}$, $\beta_{2,1}$, . . . in the different equations. Thus, in multinomial regression, the coefficient of a given predictor X is allowed to be different for each relative risk.

An intuitive way to think about multinomial logistic regression is to imagine the model as a set of binary logistic regressions. Each imagined binary logistic regression is conducted on a subsample of the data, and this subsample contains only those individuals who are either in the reference category or in the category that is being compared to the reference in that model equation.

Formally, multinomial logistic regression assumes that the outcome follows a *multinomial distribution*, conditional on the predictors included in the model. In this book, I don't discuss the multinomial distribution in detail. Essentially, the assumption is that each outcome category has a fixed probability conditional on the set of predictors included in the model and that the distribution of the number of times each category occurs in a random sample follows a binomial distribution.[vi] This assumption may be

---

[vi]On the binomial distribution, see Chapter 2, especially the section 'The Binomial Distribution" and Box 2.3.

violated, for example, if we have failed to include an important predictor of the outcome in our model. With a single sample at our disposal, it is difficult to evaluate the multinomial distribution assumption, but we should be mindful of it and consider whether our model is likely to contain all important predictors of our outcome, or whether there may be some that we have left out.

Predicted probabilities

The predicted probabilities of being in each outcome category can be calculated from the estimated coefficients for any desired combination of covariate values. In order to make the equations that follow easier to read, let's give the entire right-hand side of each multinomial logistic regression equation a short name. The first multinomial regression equation is

$$\log\left(\frac{p_1}{p_J}\right) = \alpha_1 + \beta_{1,1}X_{1i} + \beta_{1,2}X_{2i} + \ldots + \beta_{1k}X_{ki}$$

Let's call the right-hand side η_1 – that is,

$$\eta_1 \equiv \alpha_1 + \beta_{1,1}X_{i1} + \beta_{1,2}X_{i2} + \ldots + \beta_{1k}X_{ik}$$

where the symbol $\equiv$ means 'is defined as', and η is the Greek letter *eta*.[vii] Analogously, the right-hand side of the remaining multinomial logistic regression equations are called $\eta_2, \eta_3, \ldots, \eta_{J-1}$.

The predicted probability for the reference category J is calculated thus:

$$p_J = \frac{1}{1 + \exp(\eta_1) + \exp(\eta_2) + \ldots + \exp(\eta_{J-1})}$$

The predicted probability for the first outcome category is calculated thus:

$$p_1 = \frac{\exp(\eta_1)}{1 + \exp(\eta_1) + \exp(\eta_2) + \ldots + \exp(\eta_{J-1})}$$

[vii]It is a custom in statistics to use the symbol η to refer to the right-hand side of a regression model. Another statistical term for this right-hand side is 'linear predictor'.

And the probabilities for all other outcome categories are calculated in a way analogous to those for the first outcome category. In general, for outcome category $j \neq J$ (i.e. for all outcomes j other than the reference category J),

$$p_j = \frac{\exp(\eta_j)}{1 + \exp(\eta_1) + \exp(\eta_2) + \ldots + \exp(\eta_{J-1})}$$

It is not necessary for most social scientists to ever use these formulae to do hand calculations, since any statistical software capable of conducting multinomial regression will have a function to do this. On the other hand, it is useful to know that the predicted probabilities can be calculated from the model estimates. As we will see below, predicted probabilities can help us to understand the estimates from a model, especially when we illustrate them graphically.

Interpreting and illustrating the results from a multinomial logistic model

This section gives a simple example of how researchers might use multinomial logistic regression to address a social science research question. We will focus in particular on illustrating and interpreting the results from a model that involves a continuous as well as a categorical predictor variable.

Example research question

Kraaykamp et al. (2010) wished to investigate how cultural consumption varies with social status. Based on the sociological theory of distinction (see the section 'Example Data for Multinomial Logistic Regression'), they hypothesised that

a   people with high social status would be more likely to be highbrow or omnivore cultural consumers rather than 'pop only' or inactive, and
b   those with high social status would also be more likely to be (exclusively) highbrows than omnivores. (Recall that omnivores attend both highbrow and pop cultural events.)

To investigate these hypotheses, we will construct a multinomial logistic model predicting cultural consumption by social status, controlling for residence (rural vs urban). In reality, we would certainly wish to control for other variables, too, but we keep the example simple to make it easier to explain. Later in this chapter, the section 'Multinomial Logistic Regression in Action' will give a more elaborate example of investigating Kraaykamp et al.'s (2010) hypotheses.

In the FSDP survey, occupational status was measured via the ISEI. This variable was introduced in Chapter 2, in the section 'An Example with Multiple Predictors', and Figure 2.7 illustrates the distribution of ISEI in our sample. The ISEI status score was z-standardised for the purpose of the analysis. A higher number indicates higher occupational status. For the moment, we will assume that the relationship between ISEI and the log relative risk of all category comparisons is linear. In the section 'Multinomial Logistic Regression in Action', we will consider how to investigate non-linearity between a continuous predictor and log relative risks.

To address the question whether and how social status is related to cultural consumption, adjusting for area of residence, we might propose the following model:

Model 4.2:

$$\log\left(\widehat{RR}^{high}\big/_{omni}\right) = \alpha_1 + \beta_{1,1}z.ISEI_i + \beta_{1,2}Rural_i$$

$$\log\left(\widehat{RR}^{pop}\big/_{omni}\right) = \alpha_2 + \beta_{2,1}z.ISEI_i + \beta_{2,2}Rural_i$$

$$\log\left(\widehat{RR}^{inact}\big/_{omni}\right) = \alpha_3 + \beta_{3,1}z.ISEI_i + \beta_{3,2}Rural_i$$

The estimates from fitting this model on the FSDP data are shown in Table 4.4.

Table 4.4 Estimates from a multinomial logistic regression of cultural consumption on residence and ISEI social status score (Model 4.2)

Consumption (Reference: Omnivore)		Estimate	SE	RRR	95% CI for RRR
Highbrow	Intercept	0.179	0.090		
	ISEI	−0.022	0.087	0.98	[0.82, 1.16]
	Rural	0.387	0.266	1.47	[0.88, 2.51]
Pop only	Intercept	−2.284	0.230		
	ISEI	−0.901	0.196	0.41	[0.27, 0.59]
	Rural	0.871	0.459	2.39	[0.92, 5.71]
Inactive	Intercept	−1.274	0.149		
	ISEI	−0.854	0.134	0.43	[0.33, 0.55]
	Rural	0.643	0.351	1.90	[0.94, 3.76]

Note. The ISEI status score was z-standardised in this analysis. Confidence intervals were calculated using the profile likelihood method. ISEI = International Socio-Economic Index; *SE* = standard error; *RRR* = relative risk ratio; CI = confidence interval.

Interpreting results from a multinomial logistic model

In our interpretation of Table 4.4, let's focus on the evidence for the relationship between social status and cultural consumption, controlling for residence (urban or rural).

We can do this by looking at the three relative risk ratios and their confidence intervals. When interpreting the coefficients, recall that the ISEI status score is *z*-standardised for this analysis, and so the *RRR* is to be interpreted as the change in the relative risk associated with a 1-*SD* change in ISEI status score. Let's look at the results in detail:

- *Highbrow versus omnivore:* the estimated *RRR* for ISEI is 0.98, with a 95% CI [0.82, 1.16]. Since the *RRR* estimate is close to 1, and since the confidence interval suggests that the data are compatible with a small effect in either direction, it seems that there is not much evidence from this model and these data that social status is associated with the relative probability of being a highbrow versus an omnivore cultural consumer.
- *Pop only versus omnivore:* the estimated *RRR* is 0.41, with a 95% CI [0.27, 0.59]. Thus, the model predicts that higher social status is associated with lower relative probability of being a pop only cultural consumer, compared to being an omnivore. To be precise, a 1-*SD* difference in ISEI score predicts a 59% lower relative probability of being pop only, compared to being an omnivore. The CI includes only *RRR*s smaller than 1, suggesting that there is good evidence for this relationship.
- *Inactive versus omnivore:* the estimated *RRR* is 0.43, with a 95% CI [0.33, 0.55]. Thus, the model predicts that a higher social status is associated with a lower relative probability of being inactive versus being an omnivore.

In summary, it seems that higher social status is associated with a smaller relative probability to be either inactive or pop only, compared to being omnivore, but that social status does not much affect the relative probability of being highbrow versus omnivore. This suggests that people with higher status are more likely to attend highbrow activities than those with lower status, but that higher status is not necessarily associated with shunning pop concerts (recall that the omnivores are those that attend both highbrow events and pop concerts). So looking back at the research hypotheses that motivated this investigation (see the section 'Example Research Question'), it looks as though hypothesis (a) is supported by the data, while there is no strong evidence in favour of hypothesis (b).

Illustrating results from a multinomial logistic model

We can illustrate the results from a multinomial logistic regression by calculating and graphically illustrating the predicted probabilities of being in each outcome category for all combinations of the predictors. I have done this using the FSDP data and the results from Model 4.2. The resulting graph is shown in Figure 4.1.

This illustrates the relationship between ISEI status score and cultural consumption predicted by Model 4.2. In both rural and urban areas, the predicted probability of being either a highbrow or omnivore cultural consumer rises with

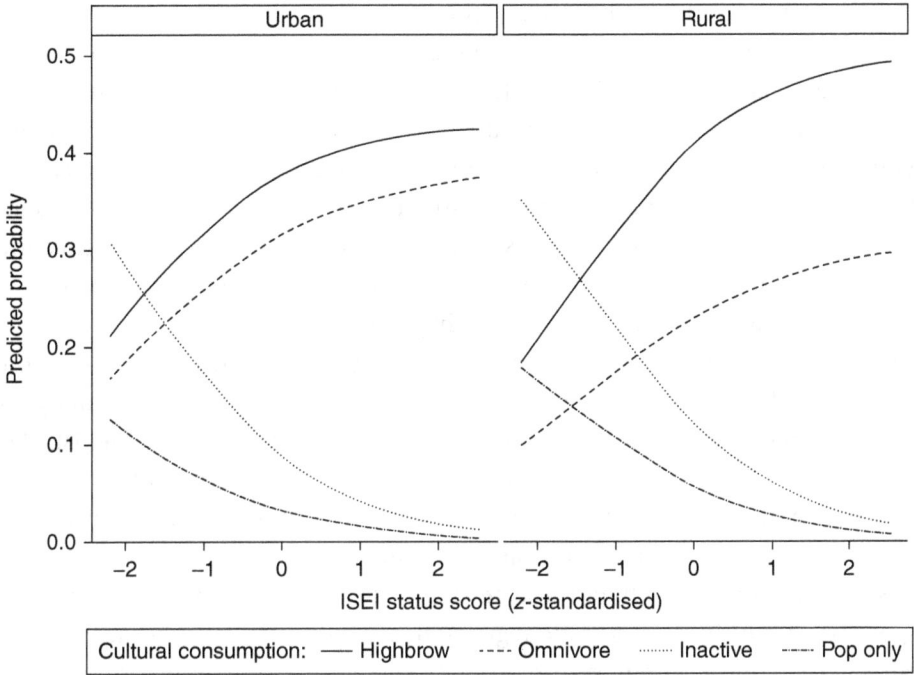

Figure 4.1 Predicted probabilities of four types of cultural consumption, contingent on residence and ISEI social status score (from Model 4.2)

Note. ISEI = International Socio-Economic Index.

occupational status, while the predicted probability of being pop only or inactive falls when occupational status rises. Furthermore, our model predicts inactives to be the most frequent category of cultural consumption among those with the lowest social status, while otherwise the most frequent category is predicted to be highbrow.

Figure 4.1 is relatively complex, even for a model with just two predictor variables and four outcome categories. You can imagine that the illustration will get more complicated still for outcomes with more than four categories, and for models with many predictors. In such cases, wise choices need to be made to devise a clear graph that illustrates the most important model results in a meaningful way. The section 'Multinomial Logistic Regression in Action' provides an example in the context of a model with six predictor variables.

Hypothesis tests and confidence intervals

Multinomial logistic regression is part of the logistic regression family of models. The most well-known and frequently used member of this family is logistic regression for binary outcomes, to which Chapter 2 of this book is devoted. The mathematical procedure for estimating the parameters of a logistic model (binary, ordinal or multinomial) is called maximum likelihood estimation (see the section 'Maximum Liklihood Estimation' in Chapter 2). This procedure gives us the best estimates of model coefficients from a given data set. It also results in a likelihood for each model, enabling us to conduct likelihood ratio tests for nested model comparisons, and to calculate profile likelihood confidence intervals. Thus, hypothesis tests and confidence intervals in multinomial logistic regression can be applied in essentially the same way as in binary and ordinal logistic regression (Chapters 2 and 3).

Nonetheless, in multinomial regression some special cases arise. This is due to the presence of multiple unordered outcome categories. We may be interested in hypotheses that specifically concern the relationship of a predictor on the relative risk of one pair of outcome categories. And we may also be interested in investigating whether two or more outcome categories may be combined without loss of information about the relationships between predictors and the outcome.

This section will discuss hypothesis testing and confidence interval estimation in the context of multinomial logistic regression. We will consider procedures that are essentially the same as in binary and ordinal logistic regression, as well as procedures that apply to multinomial models specifically.

Likelihood ratio test for comparison of nested models

When two multinomial logistic regression models are nested,[viii] we can conduct a likelihood ratio test to investigate whether there is evidence that the larger of the two models improves the prediction of the outcome relative to the smaller model. This is in principle no different to the application of likelihood ratio tests for the same purpose in the context of binary or ordinal logistic regression. However, a special feature of multinomial logistic regression is that, in general, each predictor variable is associated with a number of coefficients equal to the number of outcome categories minus one. This means that if I add one predictor variable to a multinomial logistic regression, I add several coefficients. This has implications for the degrees of freedom of the likelihood ratio test, and for the precise definition of the null and alternative hypotheses of the test. Let's consider an example to illuminate this.

Previously, in this chapter, we considered two multinomial logistic models of cultural consumption. Model 4.1 had just one predictor variable (*Rural*), while Model 4.2 had two predictors (*Rural* + *z.ISEI*). I restate both models here for convenience:

[viii]See *The SAGE Quantitative Research Kit*, Volume 7, Chapter 5 for an explanation of nested models.

Model 4.1:

$$\log\left(\widehat{RR}^{high}\!\big/_{omni}\right) = \alpha_1 + \beta_1 Rural_i$$

$$\log\left(\widehat{RR}^{pop}\!\big/_{omni}\right) = \alpha_2 + \beta_2 Rural_i$$

$$\log\left(\widehat{RR}^{inact}\!\big/_{omni}\right) = \alpha_3 + \beta_3 Rural_i$$

Model 4.2:

$$\log\left(\widehat{RR}^{high}\!\big/_{omni}\right) = \alpha_1 + \beta_{1,1} z.ISEI_i + \beta_{1,2} Rural_i$$

$$\log\left(\widehat{RR}^{uni}\!\big/_{omni}\right) = \alpha_2 + \beta_{2,1} z.ISEI_i + \beta_{2,2} Rural_i$$

$$\log\left(\widehat{RR}^{inact}\!\big/_{omni}\right) = \alpha_3 + \beta_{3,1} z.ISEI_i + \beta_{3,2} Rural_i$$

These models are nested, since I can turn Model 4.2 into Model 4.1 by setting $\beta_{1,1} = \beta_{2,1} = \beta_{3,1} = 0$. Now, we may be interested in investigating whether social status (z. ISEI) is associated with our outcome, cultural consumption, when adjusting for Rural. Then we can formulate the hypotheses of a likelihood ratio test of Model 4.2 versus Model 4.1 as follows:

- *Null hypothesis:* Model 4.2 predicts cultural consumption no better than Model 4.1. Another way of stating this is that social status (z.ISEI) is hypothesised to have no relationship with the relative risk of any pair of outcome categories. Yet another way of stating this hypothesis is that all slope coefficients associated with z.ISEI are zero – that is $H_0 = \beta_{1,1} = \beta_{2,1} = \beta_{3,1} = 0$.
- *Alternative hypothesis:* Model 4.2 predicts cultural consumption better than Model 4.1. Another way of saying the same thing is that social status is related with the relative risk of *at least one* pair of outcome categories. In other words, the alternative hypothesis is that *at least one* slope coefficient associated with z.ISEI is *not* zero. In mathematical notation, we might write the alternative hypothesis as H_1: $\beta_{j,1} \neq 0$, for at least one $j = \{1,2,3\}$.

Conducting the likelihood ratio test of Model 4.2 versus Model 4.1 on the FSDP data yields the results presented in Table 4.5.

Table 4.5 Likelihood ratio test of Model 4.2 versus Model 4.1

		LL	Λ	Parameters	df	df Difference	p
Model 4.1	*Rural*	−823.76		6	2145		
Model 4.2	*Rural + ISEI.Z*	−789.11	69.31	9	2142	3	<0.001

Note. The degrees of freedom for each model are equal to $df = n \times (J - 1) - m$, where n is the sample size, J is the number of categories of the nominal outcome and m is the number of parameters. ISEI = International Socio-Economic Index; LL = log likelihood; df = degrees of freedom.

The test statistic is $\Lambda = 69.31$, with 3 *df*, which yields a very small *p*-value ($p < 0.001$). This constitutes strong evidence against the null hypothesis, suggesting that social status is related to cultural consumption, controlling for area of residence.

I will add a few formal words about this type of likelihood ratio test for multinomial logistic regression. The test statistic has three degrees of freedom in this case, because three is the difference in the numbers of parameters between the two models. Model 4.1 has six parameters (three intercepts and three slope coefficients), Model 4.2 has nine (three intercepts and six slopes).

In general, if two nested multinomial regression models differ by a single predictor, then a likelihood ratio test comparing these two models has degrees of freedom equal to the number of outcome categories minus one ($J - 1$). Of course, it is also possible to conduct likelihood ratio tests comparing two nested models that differ in more than one predictor. In that case, the degrees of freedom for the test are equal to $(J - 1) \times q$, where q is the number of predictors present in the larger model but absent in the smaller model.

Turning back to our example, the result of the likelihood ratio test of Model 4.2 versus Model 4.1 suggests that social status is related to cultural consumption, controlling for area of residence. It does not tell us, however, *how* ISEI is related to cultural consumption: is ISEI related to the relative risks of all pairs of outcome categories or only to some? Which outcome categories are associated with a higher social status and which with a lower status? To investigate these questions, we might interpret the estimated *RRR*s and illustrate the predicted probabilities, as we have done earlier in the section 'Interpreting and Illustrating the Results From a Multinomial Logistic Model'. We might also be interested in conducting formal statistical tests of individual coefficients. This is the issue to which we turn next.

The z-test for an individual coefficient

Once we have concluded that a predictor variable has a relationship with our nominal outcome, we may be interested in investigating more specific questions about this relationship. For example, we might be interested in the specific question whether social status is related to the relative risk of being a highbrow cultural consumer versus being an omnivore. Is a higher social status associated with a greater relative probability of attending exclusively highbrow cultural activities, as opposed to being omnivorous and attending all kinds of cultural activities? This is one of the questions Kraaykamp et al. (2010) asked in their study (see hypothesis [b] in the section 'Example Research Question').

To investigate such a question, we may wish to test hypotheses about individual coefficients. For example, in Model 4.2, we may wish to test the null hypothesis that the slope coefficient of ISEI for predicting the log relative risk of being highbrow versus omnivorous is zero (i.e. H_0: $\beta_{1,1} = 0$, using the coefficient name from Model 4.2 as defined above).

Such a test can in principle be conducted as a likelihood ratio test. However, at the time of writing, such a likelihood ratio test for a single coefficient within a multinomial logistic regression is not implemented in standard statistical software in a way that would make it easily accessible to non-experts. In practice, researchers often instead use a z-test, and this is the test that I will introduce in this section.

The test statistic for a z-test about an individual coefficient in a multinomial logistic regression model is calculated as follows:

$$z = \frac{\hat{\beta} - \beta_0}{\hat{\sigma}_{\hat{\beta}}}$$

where $\hat{\beta}$ is the estimated coefficient, $\hat{\sigma}_{\hat{\beta}}$ is its estimated standard error, and β_0 is the value of the coefficient under the null hypothesis (the 'test value'). Most social scientists tend to investigate null hypotheses of the type where $\beta_0 = 0$, so that the formula becomes simply $z = \frac{\hat{\beta}}{\hat{\sigma}_{\hat{\beta}}}$. This z-test of H_0: $\beta = 0$ is usually displayed by default in multinomial regression output from statistical software. Note also that this z-test is identical in form to the Wald test of a coefficient in a binary logistic regression (see Chapter 2, section 'Wald Test').

Under the null hypothesis, the z-statistic is approximately normally distributed for relative risks estimated on large samples. When I say 'large samples', however, we need to be precise: what matters here is not the size of the whole data set but the number of cases in the two outcome categories whose comparison the coefficient estimate $\hat{\beta}$ is based on. If one or both of these groups are small, the z-test may not give reliable results and it may be better not to use it. This consideration is analogous to the consideration for the Wald test in binary logistic regression.

Let's look at an example application of the z-test. To test the null hypothesis that the coefficient for ISEI in the comparison highbrow versus omnivore is equal to zero, we use the estimates of the relevant coefficient and its standard error from Table 4.4. We have the coefficient estimate $\hat{\beta}_{1,1} = -0.022$ and the standard error estimate $\hat{\sigma}_{\hat{\beta}_{1,1}} = 0.087$. The z-statistic is then calculated as follows:

$$z = -\frac{\hat{\beta}_{1,1}}{\hat{\sigma}_{\hat{\beta}_{1,1}}} = \frac{-0.022}{0.087} = -0.25$$

This yields a two-sided p-value of $p = 0.803$, so there is little evidence from these data to suggest that ISEI is related to the choice between omnivorous and exclusively highbrow cultural consumption, once area of residence is taken into account.

To give another example, suppose we are interested in the question whether social status is related to the relative risk of being a pop only cultural consumer versus an omnivore. Then we might conduct a z-test of the null hypothesis that the coefficient of ISEI for the log relative risk of pop only versus omnivore is equal to zero. Taking again the estimates displayed in Table 4.4, we have:

$$z = -\frac{\hat{\beta}_{2,1}}{\hat{\sigma}_{\hat{\beta}_{2,1}}} = \frac{-0.901}{0.196} = -4.59$$

This yields a two-sided p-value of $p = 0.000004$, suggesting that there is strong evidence from these data that social status is related to the relative risk of pop only versus omnivore.

Confidence intervals

Confidence intervals for individual coefficients can also be calculated using the assumption that coefficient estimates are approximately normally distributed. Again, this is a reasonable assumption if both outcome categories involved in the comparison in question have large numbers of cases. A 95% CI for a coefficient is calculated thus:

$$CI_{0.95}(\beta) = \hat{\beta} \pm 1.96 \times \hat{\sigma}_{\hat{\beta}}$$

where $\hat{\beta}$ and $\hat{\sigma}_{\hat{\beta}}$ are the coefficient estimate and its estimated standard error, as defined in the previous section for the z-test. A 95% CI for the corresponding *RRR* can be calculated by exponentiating the confidence limits for the coefficient.

For example, a 95% CI for the coefficient of ISEI for the comparison pop only versus omnivore can be calculated thus:

$$CI_{0.95}(\beta_{2,1}) = -0.901 \pm 1.96 \times 0.196 = [-1.285, -0.516]$$

The confidence limits for the *RRR* are then found by calculating exp (−1.285) = 0.28 and exp (−0.516) = 0.60, respectively. Most statistical software will calculate and display such CIs by default, either for the coefficient or for the *RRR*.

Analogous to binary and ordinal logistic regression, an alternative (and sometimes better) approach to calculating confidence intervals is the profile likelihood method (see Chapter 2, section 'Confidence Intervals'). While mathematically more difficult than the normal approximation described above, profile likelihood confidence intervals are readily calculated by statistical software and have the advantage that they are more likely to be accurate than normal approximation confidence intervals when the sample size is small. In large samples, the two methods usually yield very similar results. For example, Table 4.4 displays profile likelihood confidence intervals. The profile likelihood 95% CI for the *RRR* of ISEI in the comparison pop only versus omnivore is [0.27, 0.59], which is almost the same as the one found via the normal approximation above.

Test for combining outcome categories

It may sometimes be of interest to investigate whether two or more outcome categories can be combined to simplify a multinomial logistic model. For example, suppose that before seeing the data, our hypothesis had been that 'highbrows' and 'omnivores' are really the same category of cultural consumer, in the sense that all interesting predictor variables are unrelated to the choice between being highbrow versus omnivorous.

Formally, we can investigate this question by testing the null hypothesis that all slope coefficients relating to the comparison 'highbrow' versus 'omnivore' equal zero in the population. This null hypothesis model is nested within the model that estimates these coefficients and allows them to be different from zero. Since the two models are nested, we can perform a likelihood ratio test.

Let's stay with the example I have just given and test the null hypothesis that in Model 4.2 the coefficients of both *ISEI* and *Rural* are zero in the prediction of the log relative risk of highbrows versus omnivores. Expressing this mathematically, H_0: $\beta_{1,1} = \beta_{1,2} = 0$, where $\beta_{1,1}$ and $\beta_{1,2}$ are the coefficients associated with the predictors *Rural* and *ISEI*, respectively, for the comparison highbrow versus omnivore.

Since the null hypothesis model features two fewer coefficients than Model 4.2, the likelihood ratio test comparing these models has two *df*. Conducting this test on our data, we obtain $\Lambda = 2.26$, $df = 2$, and $p = 0.323$. Thus there is no strong evidence, from these data, that either social status or area of residence (rural or urban) affect the relative probability of being a highbrow versus an omnivore cultural consumer. This suggests that we could combine the two categories without losing information about the influence of the predictors on our outcome. If the likelihood ratio test had yielded a smaller *p*-value, this would have suggested that combining the categories results in loss of information about differences in the relationships between the predictors and the outcome.

When interpreting the results from a likelihood ratio test for combining outcome categories, there are two things to bear in mind:

1   The likelihood ratio test for combining categories does not test the intercept. Thus, the test may yield a large p-value, even if the two categories have quite different frequencies in the population. In our example, 'highbrow' is a more frequent type of consumption than 'omnivore'. This distinction would get lost in the modelling if we combined the two categories. Whether this is a loss that matters depends on the aims of our particular study.

2   The result of the likelihood ratio test always pertains to a particular model. The test says nothing about a model of the same outcome but with a different set of predictors. See the section 'Multinomial Logistic Regression in Action' for another example of the likelihood ratio test for combining categories, where we are testing whether we can combine highbrows and omnivores in the context of a multinomial logistic regression with six predictors. As you will see, the result suggests a different conclusion in that case.

The result of a significance test for combining categories should not be the only, or even the main, consideration that informs the decision whether to combine outcome categories or not. As always in statistical modelling, what matters most is the suitability of the model for investigating the research question of interest. It is not necessary to conduct this test for every multinomial logistic regression you might estimate, and it is certainly neither necessary nor advisable to conduct this test for all possible pairs of outcome categories.

Multinomial regression: some additional comments

Many of the practical issues we face when using multinomial logistic regression are similar to those we encounter in binary logistic regression. But some issues are different. In this section, I discuss the choice of reference category for the multinomial outcome and things to be aware of when using dummy variables for categorical predictors in multinomial models.

How to choose the reference outcome category

From a mathematical point of view, which outcome category I choose to make the reference is arbitrary and unimportant. Both the model likelihood and the predicted probabilities (for all categories) are unaffected by the choice of reference category. However, the specific relative risks to be predicted and thus the coefficients to be estimated do depend on the reference category. Therefore a particular reference category may be more or less convenient in the context of a given investigation. For example, imagine that we wished to analyse cultural consumption in the FSDP data and our main research question was 'Are urban people more likely to be culturally active, with either highbrow or lowbrow cultural activities, than rural people'? In this case, it might have been convenient to define 'inactive' as the reference category, because this would have resulted in a model that explicitly compares each of the other cultural consumption categories to the inactives.

A second consideration in the choice of reference category is group size – that is, the number of cases placed in each outcome category in our sample data. It is often inconvenient to choose as the reference a category that has small case numbers, because this choice will mean that standard errors for the estimated slope coefficients will be relatively large. For example, in the FSDP data, there are only 38 cases in the 'pop only' category (see Table 4.2). It would probably be unwise to make this the reference category, unless there was a strong theoretical reason to do so.

Categorical predictors with dummy variables

Multinomial regression differs from many other methods covered in this book, in that every predictor variable is associated with not one coefficient but several coefficients.[ix] As we saw above, the number of coefficients estimated for each predictor is equal to $J - 1$, where J is the number of outcome categories. Now consider what happens if I investigate the relationship between a multinomial outcome and a categorical predictor with several categories. For example, Kraaykamp et al. (2010) were interested in the relationship between cultural consumption and occupational social class. In their data set, occupational social class is a categorical variable with seven categories (higher professional, lower professional, routine non-manual, self-employed/farmers, supervising manual, skilled manual and unskilled manual). To include this variable as a predictor, we would have to recode it into six dummy variables. This would mean that the relationship between social class and cultural consumption would be modelled by $3 \times 6 = 18$ coefficients. There is nothing wrong with this in principle, but in practice you may encounter one or more of the following problems:

- *Small or zero cell counts:* in a given sample, the number of cases that have a particular combination of predictor and outcome category may be small, or even zero. This is the problem of sparseness which we already discussed in Chapter 2 in the section 'Things That Might Go Wrong', in the context of binary logistic regression. For example, there may be very few 'pop only' respondents in a particular social class, or even none at all. This may cause the estimated coefficients to be very uncertain (leading to wide confidence intervals), or may even cause the model estimation to fail (such that statistical software might return an error or warning).

[ix]In the non-proportional odds model for ordinal outcomes, this is also the case. In the partial proportional odds model, it is the case for some predictors (see Chapter 3). Chapter 5 discusses zero-inflated and hurdle models, which are composed of two model parts, each of which may feature a coefficient for the same predictor.

- *Interpretation:* moreover, having a large model with many coefficients may make the results difficult to interpret, simply because there are many numbers to consider.

If you encounter problems with sparseness or interpretability, you might try one or several of the following things:

- Combine two or more outcome categories into a single category, thus reducing the number of coefficients to be estimated
- Combine two or more predictor categories into a single category, thus reducing the number of dummy variables and hence the number of coefficients to be estimated
- Consider whether the concept measured by your categorical predictor can be represented by a continuous predictor instead (e.g. instead of occupational social class categories, consider whether you can use the continuous ISEI occupational status score).

Each of these strategies may lead to some loss of information or loss of a match between the theory you wish to investigate and the model you are estimating. But they may be preferable to having an uninterpretable or poorly fitted model.

Multinomial logistic regression in action

This section will walk through a complete example of applying multinomial logistic regression. We will investigate the relationship between occupational social status and choices in cultural consumption. Our example data, as previously in this chapter, come from the FSDP, conducted in the Netherlands in 1998. The analysis is inspired by the study by Kraaykamp et al. (2010), although I have simplified it in several respects, in order to make the example easier to explain and enable us to focus on the methodological issues. Kraaykamp et al. were interested in the relationship between social status and cultural consumption, because of their theoretical interest in the sociology of social distinction. They proposed that social status is expressed in cultural tastes. Individuals will have a tendency to express, and thereby display and confirm, their social status through their cultural choices.

Put simply, a university professor may go to a classical concert not only because she enjoys the music but also to display, to herself and her peers, that she is the kind of person who appreciates high culture. Conversely, a judge may decide not to go to a pop concert because he may feel that it is not appropriate for him to be seen at an event considered culturally lowbrow. I hasten to add that my description of this theory rather oversimplifies it – Kraaykamp et al. (2010) give a more thorough account.

Kraaykamp et al. considered that the sociological theory of cultural taste would make the following predictions about the relationship between social status and cultural consumption:

- People with higher social status are more likely to be exclusively highbrow in their cultural consumption choice, rather than being omnivores who pursue both highbrow and lowbrow cultural activities.
- People with higher social status are more likely to pursue highbrow cultural activities than people with lower social status, both as 'exclusive highbrows' and as 'omnivores'.
- These differences should not be entirely explainable by other characteristics associated with both social status and cultural consumption, such as age, income, education, area of residence and so forth. For example, highbrow cultural activities might be expensive and hence more affordable to people with higher social status, who tend to have higher incomes. The theory of cultural taste would predict that two people who have the same income, but who have different social status, would be likely to make different cultural consumption choices.

In order to investigate whether these predictions are borne out by the FSDP data, we propose a multinomial logistic regression of cultural consumption on the ISEI occupational status score. We control for relevant covariates in order to reduce the risk that the relationship we estimate is biased due to confounding. In this example, we will control for income, age, education, the presence of young children in the household, and area of residence (rural or urban). (In Kraaykamp et al.'s, 2010, analysis, the number of covariates in the model was much larger.) We don't have a reason to assume that there are interactions between any of the predictors, so our model will contain main effects only.

In preparation for specifying the model, we further wish to investigate whether relationships between the numeric predictors and the log relative risks of the outcome are likely to be linear. We have three numeric predictor variables in our model: age, income and ISEI social status. I will show the exploration in detail for ISEI social status only, but thorough modelling would involve considering the question of linearity for all numeric predictors.

To explore the shape of the relationship between ISEI and cultural consumption, I categorised ISEI into five quintiles, such that the first quintile represents the 20% of sample members with the lowest social status score and the fifth quintile represents the 20% of sample members with the highest social status score. I then made a bar chart of cultural consumption by ISEI quintile group, which is shown in Figure 4.2. Table 4.6 displays the data used to make the bar chart.

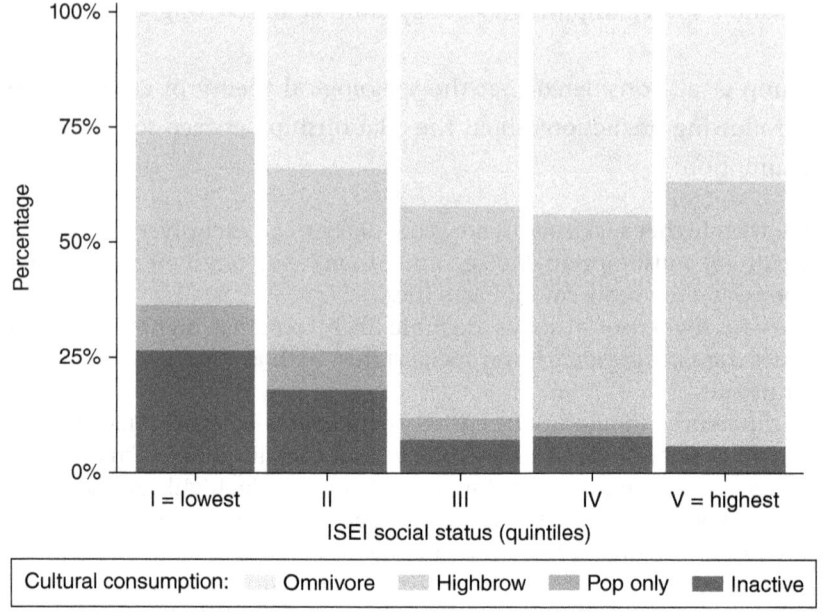

Figure 4.2 Percentages of four types of cultural consumption in the Family Survey Dutch Population, by social status

Note. ISEI = International Socio-Economic Index.

Table 4.6 Numbers (and percentages) of cultural consumption categories by ISEI social status quintiles

Cultural Consumption	ISEI Occupational Social Status Score (Quintiles)											
	I = lowest		II		III		IV		V = Highest		Total	
Omnivore	42	26%	47	34%	57	42%	61	44%	52	37%	259	36%
Highbrow	61	38%	55	40%	62	46%	63	45%	82	58%	323	45%
Pop only	16	10%	12	9%	6	4%	4	3%	0	0%	38	5%
Inactive	43	27%	25	18%	10	7%	11	8%	8	6%	97	14%
Total	162	100%	139	100%	135	100%	139	100%	142	100%	717	100%

Note. Data were taken from the Family Survey Dutch Population. ISEI = International Socio-Economic Index.

Consider Figure 4.2 and Table 4.6. We can make the following observations:

- The higher the social status, the smaller the proportion of 'inactives'. This is a clear pattern with just a small exception in ISEI quintile groups III and IV, which have approximately the same proportions of inactives.
- The higher the social status, the smaller the proportion of 'pop only' cultural consumers. This is a clear pattern.

- The higher the social status, the larger the proportion of 'highbrows'. Again, there is a small exception to this pattern in ISEI quintile groups III and IV, which have approximately the same proportions of highbrows.
- The pattern for omnivores stands out. The proportion of omnivores rises with higher social status when considering quintile groups I to IV, but group V does not conform to this pattern, having a proportion of omnivores somewhere between those of groups II and III. We have, therefore, a somewhat inverted U-shaped pattern for the relationship between social status and the proportion of omnivores.

Consider what these observations imply for the relative risks for pairs of categories. Let's take, for example, the relative risk of highbrows versus inactives. This relative risk is very clearly the larger, the higher the social status. The same can be said for the relative risks of highbrows versus pop only respondents, omnivores versus inactives, and omnivores versus pop only respondents. However, the relative risk of highbrows versus omnivores follows a U shape: it is about 1.5 for quintile group I, then gets successively smaller as we move from quintile group II to III and IV, but then is larger again (about 1.6) for quintile group V.[x]

On balance, there would be little reason to suspect that a linear term for ISEI was inappropriate for our model, except for the indication of a U-shaped relationship between ISEI and the relative risk of highbrows versus omnivores. A U-shaped relationship suggests a square transformation, so in our modelling we will explore whether a square term for ISEI improves the prediction of cultural consumption, controlling for other covariates.

I conducted similar explorations of the relationship of cultural consumption with income and age, respectively. I won't show these explorations in detail but briefly summarise the findings. Graphical exploration very clearly indicated a U-shaped relationship between some relative risks of cultural consumption categories and age. For example, the youngest and the oldest age groups were more likely to be inactive than groups in the middle of the age range. I have therefore made the decision to model age via a linear and a square term. Regarding income, the exploration indicated that a logarithmic transformation of income might improve the model, and so I have log transformed income and then centred it on its median (in the same way as I did in Chapter 2).

The result of these explorations is that the following model is proposed:

Model 4.3:

$$\log\left(\widehat{RR}\frac{j}{J}\right) = \alpha_j + \beta_{j,1} z.ISEI_i + \beta_{j,2} Rural_i + \beta_{j,3} Highschool_i + \beta_{j,4} Youngkids_i +$$
$$\beta_{j,5} AgeC_i + \beta_{j,6} AgeC_i^2 + \beta_{j,7} \log(income_i),$$

---

[x]How did I estimate these relative risks? I used the numbers from Table 4.6. In ISEI quintile group I, the relative risk for being highbrow versus omnivore in these data is $RR = 62/41 \approx 1.5$. The relative risks for groups II through V are, respectively, 1.2, 1.1, 1.0 and 1.6.

where

- the outcome is cultural consumption with reference category J = omnivore and further categories j = {'highbrow', 'pop only' and 'inactive'};
- $z.ISEI$ and $Rural$ are defined as previously in this chapter;
- $Highschool$ is a dummy variable coded 1 = completed at least high school education or 0 = lower than high school education;
- $Youngkids$ is a dummy variable coded 1 = lives with at least one child aged 0 to 4 years in the same household or 0 = no children aged 0 to 4 years in the household;
- $AgeC$ is mean-centred age, and $AgeC^2$ is the square of mean-centred age;
- log($income$) is the logarithm of income, centred so that 0 is the median income (5000 Dutch guilders – see also Chapter 2, section 'An Example With Multiple Predictors').

We estimate this model on the FSDP data. To investigate whether the addition of a square term for ISEI would improve the model, we conduct a likelihood ratio test comparing Model 4.3 with a larger model that is identical, except for the additional presence of $z.ISEI^2$. The test yielded Λ = 7.071, df = 3, and p = 0.070. This result suggests that there is some evidence that the addition of $z.ISEI^2$ improves the prediction of cultural consumption, but the evidence is not very strong. If we were using the conventional decision criterion of $p < 0.05$, for example, we would conclude that Model 4.3 is to be preferred. In my opinion, this is a borderline case where either model could be justified. For the sake of this example, I have chosen to go with the simpler model (Model 4.3).

The estimates for Model 4.3 are displayed in Table 4.7. Note that I have chosen to display z-tests for the slope coefficients of ISEI only, because these coefficients are important for evaluating the evidence regarding the research hypotheses. It is not necessary or useful to display a z-test for every coefficient estimated by the model.

To investigate the first hypothesis – that people of higher status are more likely to exclusively pursue highbrow activities – we consider the comparison of highbrow versus omnivore for the ISEI score. The estimated RRR is close to 1, with a 95% CI [0.74, 1.15]. A z-test of the null hypothesis that the RRR is equal to 1 yields p = 0.461. So we don't have good evidence to justify rejecting this null hypothesis. In other words, the idea that higher status people prefer not to be seen at 'lowbrow' pop concerts is not supported by this analysis. In fact, the point estimate of the RRR is 0.92, which appears to indicate that a higher social status reduces the relative risk of being exclusively highbrow compared to omnivore, once we control for all the other variables in the model. This suggests that, if anything, the relationship is in the opposite direction to the one hypothesised.

To investigate the hypothesis that people of higher status are more likely to pursue highbrow activities (whether exclusively or as 'omnivores', i.e. attending both highbrow activities and pop concerts), we look at the comparisons pop only vs omnivores

Table 4.7 Estimates from a multinomial logistic regression model predicting cultural consumption (Model 4.3)

Cultural Consumption (vs. Omnivore)		Estimate	SE	RRR	95% CI for RRR	z	p
Highbrow	Intercept	−0.343	0.160				
	ISEI	−0.084	0.114	0.92	[0.74, 1.15]	−0.74	0.461
	Rural	0.521	0.291	1.69	[0.95, 2.98]		
	Highschool	0.349	0.215	1.42	[0.93, 2.16]		
	Youngkids	0.456	0.229	1.58	[1.01, 2.47]		
	AgeC	0.095	0.012	1.10	[1.07, 1.13]		
	AgeC²	0.002	0.001	1.002	[1.000, 1.004]		
	log(income)	−0.328	0.269	0.72	[0.43, 1.22]		
Pop only	Intercept	−2.263	0.343				
	ISEI	−0.683	0.232	0.51	[0.32, 0.80]	−2.95	0.003
	Rural	0.722	0.471	2.06	[0.82, 5.18]		
	Highschool	−0.319	0.555	0.73	[0.25, 2.16]		
	Youngkids	−0.030	0.439	0.97	[0.41, 2.30]		
	AgeC	−0.006	0.021	0.99	[0.95, 1.04]		
	AgeC²	0.000	0.002	1.000	[0.997, 1.004]		
	log(income)	−1.488	0.584	0.23	[0.07, 0.71]		
Inactive	Intercept	−1.836	0.246				
	ISEI	−0.540	0.163	0.58	[0.42, 0.80]	−3.31	0.0009
	Rural	0.589	0.372	1.80	[0.87, 3.74]		
	Highschool	−0.337	0.389	0.71	[0.33, 1.53]		
	Youngkids	1.031	0.324	2.80	[1.49, 5.29]		
	AgeC	0.062	0.016	1.06	[1.03, 1.10]		
	AgeC²	0.003	0.001	1.003	[1.001, 1.005]		
	log(income)	−1.828	0.414	0.16	[0.07, 0.36]		

Note. ISEI was z-standardised. AgeC is measured in years and mean centred. AgCe² is the square of mean-centred age. log(income) is the logarithm of income, centred at the median. Statistical tests: p-values are shown for z-tests of the null hypothesis that the coefficients of ISEI are equal to zero. 95% CIs were calculated by normal approximation. SE = standard error; RRR = relative risk ratio; CI = confidence interval; ISEI = International Socio-Economic Index.

and inactive versus omnivores. In both cases, the estimated RRR is well below 1, and the p-values of the z-tests suggest that we have strong reason to reject the null hypothesis of RRR = 1 in both cases. Thus, our second research hypothesis is confirmed: higher occupational status is associated with a tendency to pursue highbrow activities, controlling for the other variables in the model. Specifically, a person with an ISEI 1 SD higher than another person is predicted to have a 49% lower relative probability to attend pop concerts only, compared to being omnivorous (since the estimated *RRR* is 0.51). Similarly, a 1 *SD* difference in ISEI is associated with a 42% lower relative probability to be inactive rather than omnivorous (the estimated *RRR* is 0.58).

We might also be interested in the comparisons highbrow versus pop only and highbrow versus inactive. We can already see from our results that higher ISEI is associated with higher probability of being 'highbrow' compared to those categories: since the *RRR* of highbrow versus omnivore is about 1, the *RRR*s of highbrow versus pop only and highbrow versus inactive are pretty much the same as the *RRR*s of omnivore versus pop only and omnivore versus inactive, respectively. If we wanted to confirm this formally, and obtain confidence intervals and *z*-tests for these comparisons, we could change the reference category to 'highbrow' and re-estimate the model.

I wish to give an illustrative example of an application of the likelihood ratio test for combining outcome categories (see the section 'Test for Combining Outcome Categories'), although this test is not essential to answering the research question in this particular case. For the sake of example, let's imagine that we are interested in the exploratory question whether there is any difference in the demographic characteristics of highbrows versus omnivores at all. In order to investigate this, we can use the likelihood ratio test for combining outcome categories, investigating the null hypothesis that all the *RRR*s for the comparison highbrow versus omnivore are equal to 1. This test yields $\Lambda = 132.12$, $df = 7$ and $p < 0.001$. Thus, although social status may not be related to the choice between highbrow and omnivore, at least one other predictor in this model appears to be. From Table 4.7, it seems that an important predictor for this distinction is age: the older the respondent, the more likely they are to be exclusively highbrow rather than omnivorous.

Finally, it will be useful to display the model results graphically, by plotting the predicted probabilities of the outcome categories against predictor variables of interest. In this instance, ISEI is the most interesting variable. Since I have concluded above that age may be an important predictor of the relative probability of being highbrow versus omnivore, I also include age in my graphical inspection. Figure 4.3 shows the predicted probabilities of the four cultural consumption categories by ISEI social status score, separately for four example ages. To calculate these predicted probabilities, the variables *Highschool*, *Youngkids* and area of residence are held constant at the values 'high school education', 'no young children' and 'urban', respectively. The variable *Income* is held constant at its median.

Figure 4.3 allows us to inspect the estimated relationship between social status and cultural consumption, as well as the estimated relationship between age and cultural consumption. For example, we might make the following observations:

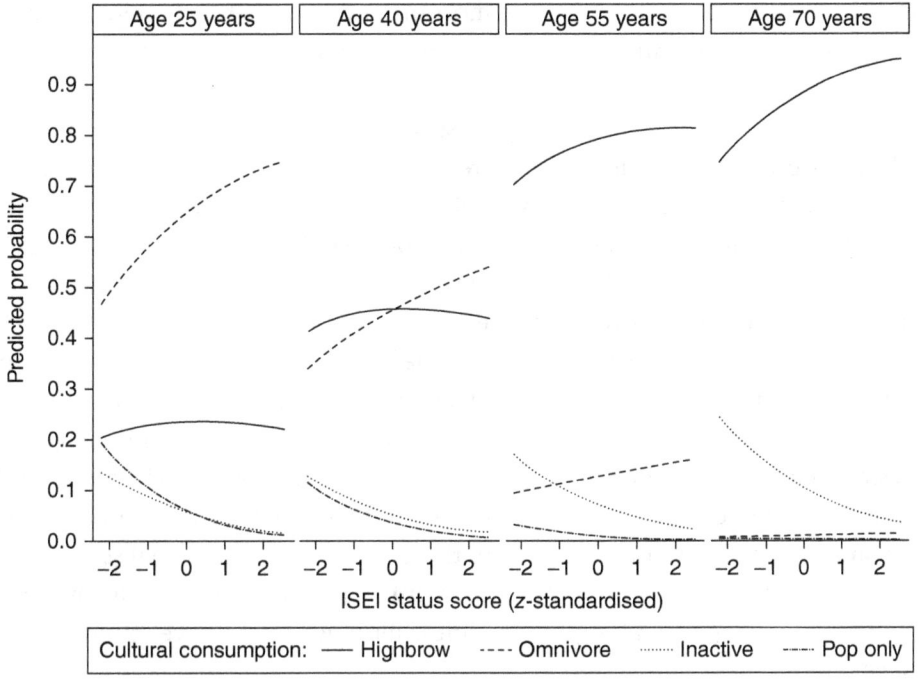

Figure 4.3 Predicted probabilities of cultural consumption categories, by ISEI status score and age, from Model 4.3

Note. Predicted probabilities are shown for people with high school education living in urban areas without young children in the household and with median income (5000 guilders per month). ISEI = International Socio-Economic Index.

- *Social status:* the probability of being an omnivore cultural consumer rise with social status, whereas the probabilities of being inactive or pop only decline with social status. This is true at all ages. The probability of being a highbrow cultural consumer tends to rise with higher social status, but at younger ages this association is very small.
- *Age:*
  - The probability of being highbrow clearly rises with age.
  - The probability of being an omnivore clearly declines with age.
  - The probability of being inactive seems to be only mildly affected by age for the age range from 25 to 40 years, but then it rises slightly with age.
  - Finally, the probability of exclusively attending pop concerts declines with age and is almost zero for those aged 70 years.

Of course, we could have chosen to construct Figure 4.3 differently, for example, by showing age on the *x*-axis, and creating separate panels for example values of ISEI or income, or by showing results separately for urban and rural respondents, and so forth. The graphical choices we make should be informed by our research interests. That is to say, we should construct a graph that allows us to investigate those aspects of the model that we are most interested in. When including a graph in a research report, we should construct the graph so that it truthfully illustrates the finding we wish to highlight.

A final word on our conclusion: with respect to the first research hypothesis, the conclusion from this analysis is that there is little evidence for the idea that among residents in the Netherlands, adults with higher social status are more likely to exclusively attend highbrow cultural activities, as opposed to being omnivore cultural consumers, when controlling for income, education, age, the presence of young children in the household and area of residence. So the idea that some people with high social status avoid pop concerts because of their status as culturally lowbrow is not supported by our model. However, our exploration of the relationship between ISEI and cultural consumption in Figure 4.2 suggested a different possibility. Figure 4.2 suggested that there may be a non-linear relationship between social status and the relative probability of being highbrow versus omnivore, such that the effect of social status on this relative risk only pertains to those with the highest social status (in the top two quintiles of the status distribution). If we had decided to retain the square term of ISEI in our model, alongside the linear term, we might have found evidence for such a result.

In the study that inspired this example, Kraaykamp et al. (2010) use a larger data set, combining data from four surveys conducted over a period of 11 years. They also control for a larger number of covariates. Their conclusions are essentially the same as those we came to in our small example analysis in this section: social status predicts higher relative probability of being omnivore or highbrow compared to being inactive or pop only, but there is little evidence that social status is associated with the relative risk of being highbrow compared to omnivorous. Kraaykamp et al. do not report exploring a non-linear model for the relationship between social status and the log relative risks of cultural consumption categories.

Chapter Summary

Multinomial logistic regression is a model for investigating how one or several predictors relate to a nominal outcome with three or more categories. If the interest is solely in a single outcome category versus all others, it may be better to construct a binary outcome variable and consider binary logistic regression. For example, if our interest in cultural consumption focuses exclusively on predictors of attending 'highbrow' cultural events, then we may not need to model the four-category outcome variable that has been the subject of the examples in this chapter. Instead, a binary variable dividing the sample into those who do and those who don't attend highbrow events may have been appropriate.

In the past, multinomial regression has sometimes been used to model ordinal outcomes in situations where the proportional odds assumption did not hold and a proportional odds model was therefore deemed inappropriate. Although this may be a justifiable choice in some instances, two reasons speak against this practice:

1   Multinomial logistic regression, if applied to an ordered outcome, ignores the ordering of outcome categories.
2   Multinomial logistic regression estimates a separate coefficient for each pairwise comparison of outcome categories, which creates an inefficient model in cases where there are many predictors, for some of whom proportional odds may be safely assumed.

It seems that in the past some researchers who encountered problems with the proportional odds assumption failed to consider a partial proportional odds model as an alternative, maybe because of a lack of familiarity with it. In general, then, when the outcome is ordinal, a generalised ordered logit model (whether assuming full, partial or no proportional odds) is likely to be more appropriate than multinomial logistic regression.

When the outcome is genuinely nominal (unorderded), an alternative to multinomial logistic regression is multinomial probit regression. This is analogous to binary probit regression, which was mentioned in Chapter 2. Agresti (2013) gives a brief definition and an example for multinomial probit regression.

Further Reading

Agresti, A. (2013). *Categorical data analysis* (3rd ed.). Wiley.

Hilbe, J. M. (2009). *Logistic regression models*. Chapman & Hall.

I recommend these two books for more mathematical and technical details on multinomial.

Liao, T. F. (1994). *Interpreting probability models: Logit, probit, and other generalized linear models*. Sage.

Liao's book has a useful focus on interpretation.

Long, J. S., & Freese, J. (2014). *Regression models for categorical dependent variables using Stata* (3rd ed.). Stata Press.

This book os you detailed guidance on how to conduct a multinomial logistic regression in Stata.

5

REGRESSION MODELS FOR COUNT DATA

Chapter Overview

This chapter is concerned with models that are suitable when the outcome variable represents something we can count. Count variables occur frequently in social research. Some examples are:

- The number of crimes committed in a city district per year
- The number of HIV tests conducted in a medical centre per month
- The number of difficulties an elderly person experiences with activities of daily living (Zaninotto & Falaschetti, 2011)
- The number of members of a parliament who switch party allegiance per year (King, 1988).

Let's state a few obvious things about count variables. Count variables are numeric but discrete (not continuous). Possible values for counts are the numbers 0, 1, 2, . . . Negative counts don't occur, since we cannot observe 'minus two crimes'. Also, counts come in whole numbers, since we cannot observe '1.3 HIV tests'. Formally, we say that count variables take as their values the *non-negative integers*.

In many cases, count variables are defined in relation to a period of time, an area of space, or a reference population. For example, if we wanted to compare city districts with respect to the number of crimes that occur in each, it would be advantageous to count the number of crimes over the same period in each district (e.g. we might define our count outcome to be the number of crimes committed in the year 2019). In cases where our units of analysis are measured over different periods of time or different sizes of areas, or if they differ in their populations, we might wish to adjust for these differences in our analysis. For example, in an analysis of crimes in city districts, we might wish to adjust for differences in the population sizes of the districts.

Linear regression is sometimes used for modelling count outcomes, but often it is not suitable, because the errors around predicted counts are often neither normally distributed nor homoscedastic (see Chapter 1 in this book for a brief review of the assumptions of linear regression). Also, linear regression can potentially result in negative predicted counts, which would be unsatisfactory. In general, count data often violate the assumptions of linear regression, and if they do, then the standard errors, confidence intervals and hypothesis tests from a linear regression of a count outcome are likely to be incorrect. So we need other models, which are specifically suited to the analysis of count outcomes.

This chapter will introduce eight types of models for count outcomes, and consider how to choose the most appropriate one for a given research problem and data set. To help us understand these models, let us first consider how count variables are typically distributed.

Distributions for count data

We begin with a famous example of a count variable. Consider Table 5.1, which shows the number of deaths by horse kick per year in Prussian army corps from 1875 to 1894. The data span 20 years for each of 14 army corps, so there are 20 × 14 = 280 observations. Table 5.1 shows that 0 deaths occurred in 144 cases, 1 death occurred in 91 cases, and so forth. Formally, we refer to the values of the variable as the *number of events* – in this case, the number of deaths.

Table 5.1 Number of deaths by horse kick per year in 14 Prussian army corps, 1875 to 1894

Number of Deaths per Year	Frequency	Percentage
0	144	51.4
1	91	32.5
2	32	11.4
3	11	3.9
4	2	0.7
Total	280	100.0
Mean	0.70	
Variance	0.76	

Note. Data were taken from von Bortkiewicz (1898) and Jackman (2017).

Two theoretical distributions are often used to describe count data: the Poisson distribution and the negative binomial distribution. Both describe the probability of observing y events, where y may be 0, 1, 2, . . . We will discuss each distribution in turn.

The Poisson distribution

The Poisson distribution is the simplest probability distribution for count data. It describes the probability that we observe 0, 1, 2, . . . events. The Poisson distribution is a theoretical distribution, like the normal distribution is one of several theoretical probability distributions for continuous variables, and the binomial distribution is a theoretical probability distribution for binary variables.

A formal definition of the Poisson distribution is given in Box 5.1. This definition requires thinking through a bit of mathematics, which some readers may like, others less so. From a practical point of view, it is important to understand the following characteristics of the Poisson distribution:

- *One parameter:* The Poisson distribution has only one parameter. We will call this parameter μ. It is equal to both the mean and the variance. Contrast this with the normal distribution, which has two parameters, the mean μ and the variance σ^2.
- *Equidispersion:* From the first characteristic follows a second: in a Poisson distribution, the mean and the variance are the same. This property is called equidispersion.
- *Positive skew:* The Poisson distribution is positively skewed: it has a longer tail to the right than to the left. The extent of the skew depends on the mean: the smaller μ, the larger the skew.

To illustrate these characteristics, Figure 5.1 shows three Poisson distributions with different means. For small μ, the Poisson distribution has a small variance and is strongly positively skewed. The larger μ, the larger the variance, and the more closely symmetric the Poisson distribution is. For $\mu \geq 10$, the Poisson distribution resembles a normal distribution, but with the important difference that the Poisson distribution only features non-negative integer values (0, 1, 2, . . .), whereas the normal distribution can include negative and non-integer values.

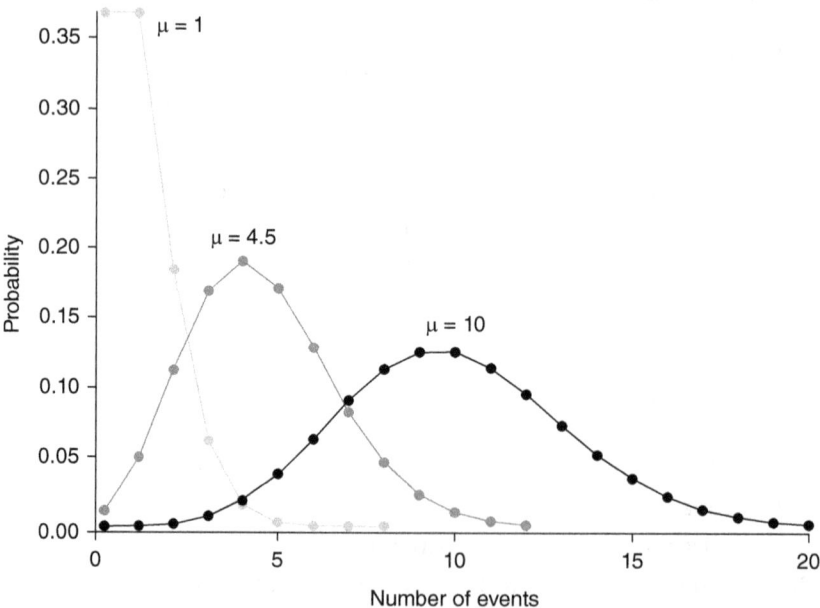

Figure 5.1 Three Poisson distributions with means equal to 1, 4.5 and 10

Note. Very small probabilities have been suppressed in this graph. For example, for the Poisson distribution with $\mu = 1$, there is a small probability to observe 9 events, 10 events, or any larger number. For all distributions shown here, there is a small probability of observing 21, 22, . . . events. But these probabilities are so small that they are not worth showing. Of course, the probability of observing negative numbers of events is zero.

Box 5.1

Poisson Distribution

The Poisson probability mass function is

$$p(y) = \frac{\mu^y e^{-\mu}}{y!}, y = 0, 1, 2, \ldots$$

This equation allows us to specify the probability, $p(y)$, of observing y events, under the assumption that our outcome follows a Poisson distribution with mean μ. The exclamation mark symbol (!) indicates a factorial, which is defined as follows: $y! = 1 \times 2 \times 3 \times \ldots \times y$. For example, $3! = 1 \times 2 \times 3 = 6$. Also, $0! = 1$ by definition.

For example, for $\mu = 0.7$, we can use the Poisson probability mass function to calculate the probability that y equals 0, 1, 2, . . . These calculations are shown in Table 5.2.

Table 5.2 Calculating the probability that y equals 0, 1, 2, 3 or 4, in a Poisson distribution with mean $\mu = 0.7$

y	$p(y) = \dfrac{\mu^y e^{-\mu}}{y!}$, for $\mu = 0.7$
0	$\dfrac{0.7^0 e^{-0.7}}{0!} = \dfrac{1 \times 0.4966}{1} = 0.4966$
1	$\dfrac{0.7^1 e^{-0.7}}{1!} = \dfrac{0.7 \times 0.4966}{1} = 0.3476$
2	$\dfrac{0.7^2 e^{-0.7}}{2!} = \dfrac{0.49 \times 0.4966}{2} = 0.1217$
3	$\dfrac{0.7^3 e^{-0.7}}{3!} = \dfrac{0.343 \times 0.4966}{6} = 0.0284$
4	$\dfrac{0.7^4 e^{-0.7}}{4!} = \dfrac{0.2401 \times 0.4966}{24} = 0.0050$
$\vdots$	$\vdots$

(Continued)

So for a Poisson distribution with mean 0.7, the probability of zero events is 0.4966, the probability of one event is 0.3476 and so forth. You can see these probabilities plotted in Figure 5.2.

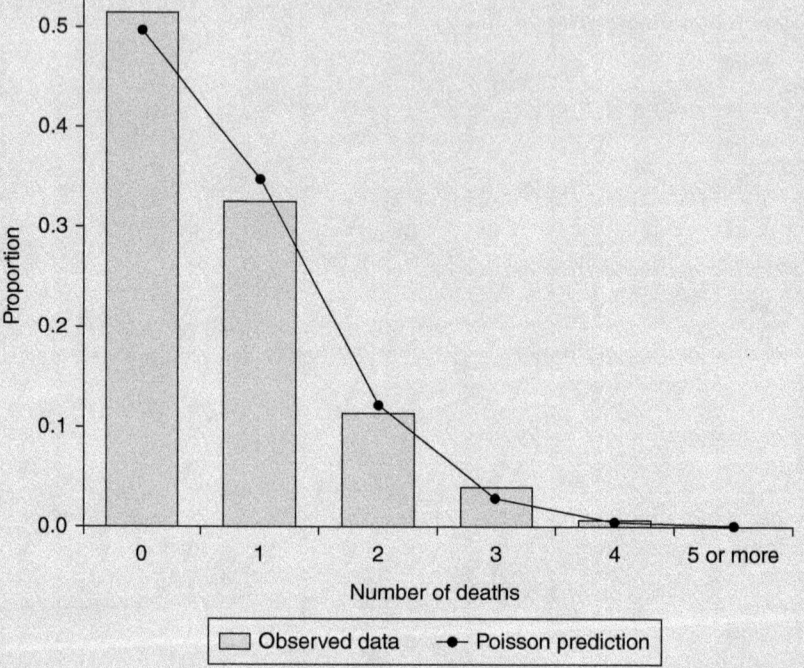

Figure 5.2 Number of deaths by horse kick: observed and expected proportions

Note. The grey bars show the proportion of times that there were 0, 1, 2, . . . deaths from horse kick. The black dots show the predicted proportions under the assumption that the data come from a Poisson distribution with the observed mean (0.7). See note to Table 5.1 for data sources.

The calculations for 5, 6 or more events are not shown in Table 5.2. However, since we know that overall the probabilities must sum to 1, we can calculate the probability of observing five or more events as follows:

$$p(y \geq 5) = 1 - \left[p(0) + p(1) + p(2) + p(3) + p(4) \right]$$

which in this case gives

$$p(y \geq 5) = 1 - \left[0.4966 + 0.3476 + 0.1217 + 0.0284 + 0.0050 \right]$$
$$= 1 - 0.999$$
$$= 0.001$$

So for a Poisson distribution with mean 0.7, there is about a 0.1% probability of observing five or more events.

A famous example of data that approximately conform to the Poisson distribution is the number of deaths by horse kick in Prussian army corps. Recall that these data record 280 counts, with observed mean $\bar{x} = 0.7$. Have a look at Table 5.3. This contrasts the observed numbers and proportions of deaths with those predicted by the Poisson distribution with $\mu = 0.7$. Contrasting the observed distribution with a theoretical distribution allows us to check how well the observations fit our statistical model. Box 5.1 explains how the probabilities (or predicted proportions) are calculated. The predicted frequencies were calculated as *predicted frequency = n × predicted proportion*, with $n = 280$ in this case.

Table 5.3 Number of deaths by horse kick: observations and Poisson predictions

Number of Deaths per Year	Observed		Predicted by Poisson Distribution	
	Frequency	Proportion	Frequency	Proportion
0	144	0.514	139.0	0.497
1	91	0.325	97.3	0.348
2	32	0.114	34.1	0.122
3	11	0.039	7.9	0.028
4	2	0.007	1.4	0.005
5 or more	0	0	0.2	0.001
Total	280	1	280	1

Note. Predicted frequencies do not always exactly equal 280 × *predicted proportion* because of rounding errors. For data sources, see the note to Table 5.1.

Table 5.3 shows that the observed frequencies are very close to the predictions. Thus the data are consistent with the idea that the observations come from a Poisson distribution. It is also often useful to make a visual comparison of observed and predicted proportions (or observed and predicted frequencies). This is shown in Figure 5.2. We again see a close match between predictions and observations.

The negative binomial distribution

We said in the previous section that in a Poisson distribution the mean and variance are the same, and that this property is called **equidispersion**. However, in social science count data, often the variance is larger than the mean, a phenomenon that is called **overdispersion**. If a variable is overdispersed, the Poisson distribution will not describe it well. An alternative in such a case is the negative binomial distribution, which allows the variance to be different from the mean.

In contrast to the Poisson distribution, which only has a single parameter μ, the negative binomial distribution has two parameters: the mean μ and a dispersion parameter, usually denoted by α (the Greek letter alpha). The dispersion parameter can take values of 0 or larger. It cannot be negative.

The variance of the negative binomial distribution is often assumed to be equal to $\mu + \alpha\mu^2$. (But other options exist.[1]) When $\alpha = 0$, then the variance is equal to the mean μ, and the negative binomial distribution is the same as the Poisson distribution. When $\alpha > 0$, the variance is larger than the mean. Figure 5.3 illustrates three negative binomial distributions that have the same mean ($\mu = 4.5$) but different dispersion parameters ($\alpha = 0$, $\alpha = 0.5$ and $\alpha = 1$, respectively). Notice that α influences both the variance and the shape of the distribution: the larger α, the larger is the variance relative to the mean, and the larger is also the positive skew.

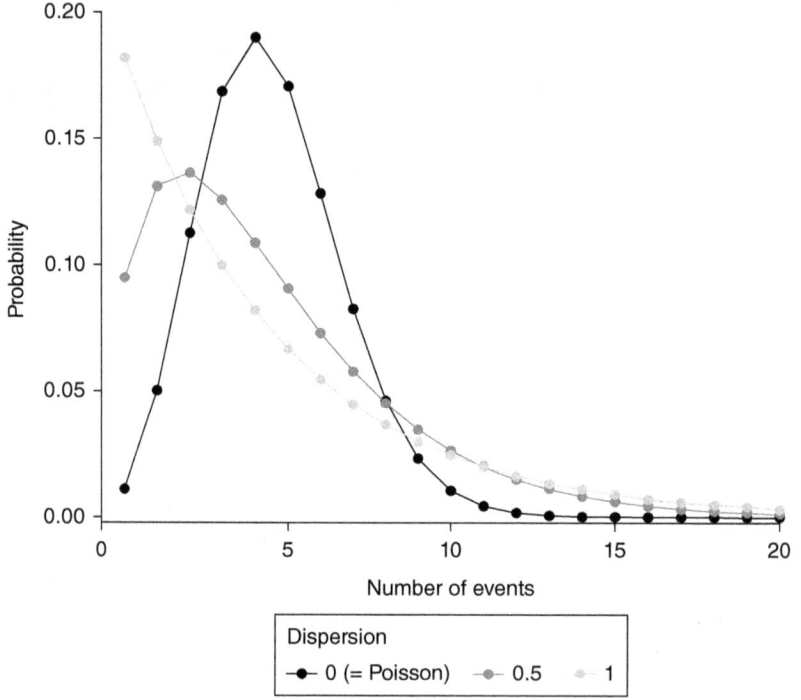

Figure 5.3 Three negative binomial distributions with mean = 4.5 and different dispersions

Now that we have considered the Poisson and negative binomial distributions, let's explore how we can use them within statistical models that investigate the relationships between a count outcome and a set of predictors. As we will see, deciding whether a count variable is equidispersed or overdispersed, and thus whether a Poisson or negative binomial distribution provides a better fit, is an important consideration when deciding which model to use for a given analysis.

---

[1]There are other ways to define the dispersion and the variance within a negative binomial model. We won't discuss them in detail in this book, but Box 5.2 gives a little extra information. For a more comprehensive discussion of the negative binomial distribution, see Hilbe (2014, Chapter 8) or Hilbe (2012).

Poisson regression

Poisson regression is appropriate when our outcome is a count variable that we can assume to follow a Poisson distribution around a predicted mean. Poisson regression is defined as follows:

$$\log(\mu_i) = \beta_0 + \beta_1 X_{1i} + \beta_2 X_{2i} + \ldots + \beta_k X_{ki}$$

$$Y_i \sim Poisson(\mu_i)$$

where the symbol ~ means 'is distributed as'. That is, we assume that our count outcome, Y_i, has a Poisson distribution with mean μ_i, where μ_i depends on the values of X_1, X_2, ... and their coefficients. The outcome is usually log transformed, since this often better represents the relationships between the outcome and the predictors, compared to a model without a transformation. The log transformation of the outcome also ensures that the model cannot result in negative predicted values.[ii]

Equivalently, we can write the Poisson model equation as

$$\mu_i = \exp(\beta_0 + \beta_1 X_{1i} + \beta_2 X_{2i} + \ldots + \beta_k X_{ki})$$

We derive this equation by exponentiating both sides of the previous equation.

The Poisson regression model looks very similar to a linear regression with a logarithmic transformation of the outcome. The difference is in our assumption about the distribution of the outcome. In linear regression, we assume that the observed outcome follows a normal distribution around the predicted mean. In Poisson regression, we assume that the observed numbers of events follow a Poisson distribution around the predicted mean. This has some interesting implications. In particular, recall the Poisson distribution's property of equidispersion: the variance is equal to the mean. This implies that in a Poisson regression we expect the observations to have a variance equal to their predicted mean. So the error variance depends on the mean, and thus by definition the prediction errors are expected to be heteroscedastic – not homoscedastic, as is assumed by linear regression.

Also, in linear regression, we can find the coefficient estimates via the method of least squares. In Poisson regression, we use the method of maximum likelihood instead, as we do for the logistic regression models discussed in Chapters 2 to 4.

[ii]It is possible to use no transformation, or a different transformation than the logarithmic, but this is rarely done in practice when choosing a Poisson regression. Consider, however, Box 5.2 for a brief discussion of different outcome transformations in the context of negative binomial regression.

Research example: police operations against street vendors in Latin American capitals

To see how Poisson regression works in practice, let's look at an example. Holland (2015) investigated why Latin American mayors sometimes choose not to take action against illegal street vendors in the cities or districts they govern. Illegal street vending allows many poor people to make an income while avoiding fees, rents and taxes, but it is unpopular among middle-class people. Holland posited that intentional non-enforcement of the laws against vendors was one way in which local politicians could benefit the poor and thus attract their votes in elections.[iii] To test her theory, she contrasted three Latin American capital cities: Lima in Peru, Santiago in Chile, and Bogotá in Colombia. Holland was interested in examining which factors predict the number of police operations against street vendors and hypothesised that these factors would be different in the three cities:

- Lima and Santiago are politically decentralised: each of the city's districts elects its own mayor, and districts differ in the relative proportions of poor and non-poor residents. Holland expected that mayors of well-off districts would have an incentive to direct police resources against street vendors, thus hoping to attract the votes of their middle-class constituents. On the other hand, mayors of poorer districts would be more likely to *not* enforce the law, thus hoping to secure the votes of the poor. So if Holland's theory is correct, the number of police operations would be expected to be higher in richer districts of Lima and Santiago, but would not necessarily be related to the number of street vendors who actually operate in a district.
- In contrast to Lima and Santiago, Bogotá is politically centralised, having one mayor for the whole city. The decision on whether to enforce the law against street vendors is made centrally rather than locally in each district. Thus, Holland expected that the mayor in Bogotá would direct resources to areas in line with the scale of the illegal activity. If this was true, the number of police operations against street vendors would be greater in districts with higher numbers of street vendors, but would not be affected by the poverty levels in the district.[iv]

To illustrate how Poisson regression can help us put Holland's theory to the test, let's first consider Bogotá. Figure 5.4 plots the number of police operations against the number of street vendors in each of Bogotá's 19 districts and also represents each district's poverty level and population size.

[iii]For example, Holland calculated that in Lima, allowing a street vendor to trade without paying rent amounts to an annual benefit for the vendor that is more than 10 times higher than the average payout to a poor family via the largest social welfare programme in the city (Holland, 2015, p. 158).

[iv]My description of Holland's study simplifies the matter in various ways. Her theory and methodology are more sophisticated than I do justice. As usual, since our focus is on the statistical methods, a little simplification serves the purpose of this book. I recommend reading Holland's article.

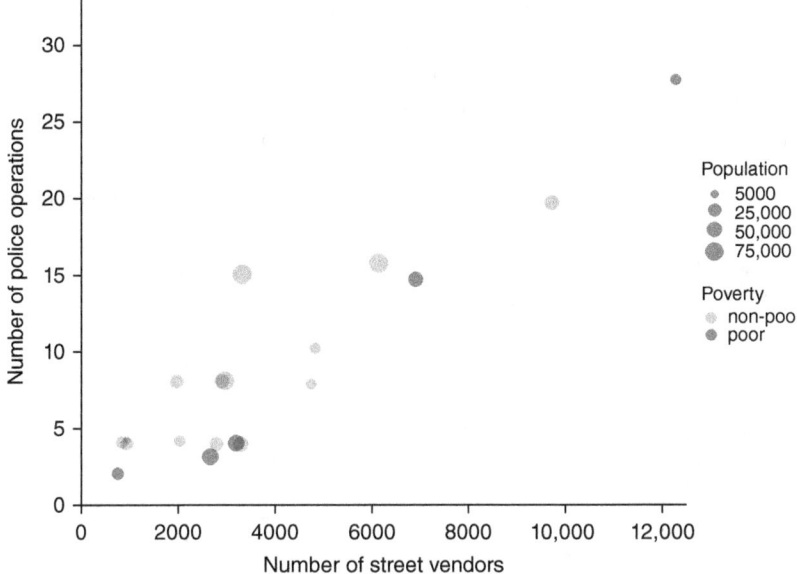

Figure 5.4 Number of police operations by number of street vendors in 19 Bogotá districts

Note. Each dot represents one city district. Dots were jittered out of position slightly to make visible multiple districts with the same number of vendors and police operations. Data provided by Alisha C. Holland via the Harvard Dataverse (Holland, 2015). 'Poor' districts are defined to be those with 50% or more lower class residents (cf. Holland, 2015).

There is a clear relationship between the number of street vendors and the number of police operations in Bogotá: the more street vendors, the more police operations against them. It's not entirely clear from this plot whether district poverty and population size are also related to the number of police operations in Bogotá.

Poisson regression of police operations in Bogotá

We can use Poisson regression to model the Bogotá data and formally investigate Holland's hypothesis. We define the model as follows:

Model 5.1:

$$\log(\mu_i) = \beta_0 + \beta_1 Vendors1000_i + \beta_2 Population1000_i + \beta_3 Lower10_i$$

$$Operations_i \sim Poisson(\mu_i)$$

where

- $i = 1, \ldots, 19$ identifies the 19 districts of Bogotá;
- $Operations_i$ denotes the number of police operations in district i;

- *Vendors1000$_i$* denotes the number of street vendors in district i (in thousands, i.e. a district with 2000 vendors has the value 2);
- *Population1000$_i$* denotes the population of district i (in thousands);
- *Lower10$_i$* denotes the percentage of the population of district i who are lower class (in tens of percentage points, i.e. a district with 20% lower class residents has the value 2)^v;
- *Operations$_i$ ~ Possion(μ_i)* – this means that the number of police operations in district i is assumed to follow a Poisson distribution, with the mean μ_i dependent on the number of vendors, the population size and the proportion of lower class residents, as specified in the model.

The estimates from fitting Model 5.1 to the Bogotá data are shown in Table 5.4.

Table 5.4 Estimates from a Poisson regression of the number of police operations in 19 Bogotá districts on number of street vendors, population size and percentage of lower class residents (Model 5.1)

	Estimated Coefficient	SE	IRR	95% CI
Intercept	1.069	0.213		
Vendors (per 1000)	0.186	0.021	1.204	[1.154, 1.256]
Population (per 1000)	0.007	0.003	1.007	[1.002, 1.012]
Percentage lower class (per 10%)	−0.014	0.024	0.986	[0.941, 1.033]

Note. Confidence intervals were calculated via normal approximation. *SE* = standard error; *IRR* = incidence rate ratio; CI = confidence interval.

The coefficients relating to the predictors are interpreted as slopes of a regression line, just like in all regression models. Since the outcome is log transformed, the coefficient estimates do not have an intuitive interpretation. The precise interpretation of a coefficient in a Poisson regression is 'the predicted difference in the log of the number of outcome events for a 1-unit difference in the predictor'. For example, the model predicts that, for every additional 1000 vendors in a district, the log of the number of police operations increases by 0.186, if the district population and the percentage of lower class residents remain constant.

Although the coefficients may not be easy to interpret in themselves, we can at least use them to interpret the direction of the estimated relationships. Thus, the estimates suggest that the number of vendors is positively related to the number of police operations, controlling for district population and proportion of lower class residents. This confirms our impression from Figure 5.4. Population size is also positively related to the number of police operations, controlling for number of vendors and proportion of lower class residents. Finally, the percentage of lower class residents is negatively associated with the number of police operations, controlling for vendors

---

^{v}The 'poor' versus 'non-poor' classification of districts in Figure 5.4 is based on the variable *Lower10* – see note to Figure 5.4.

and population. These are the results from the data set at hand. So far, we have not considered inference, such as confidence intervals or statistical tests, about the coefficients; these are explained in the section 'Model Comparison and Inference'.

The incidence rate ratio

A measure of effect that allows a more intuitive interpretation of the results from a Poisson regression is the **incidence rate ratio** (*IRR*). It is also sometimes simply called the **rate ratio**. It can be calculated from a coefficient by exponentiation. Thus, for example, the incidence rate ratio for vendors is calculated as follows:

$$\widehat{IRR}_{vendors1000} = \exp\left(\hat{\beta}_{vendors1000}\right) = \exp(0.168) = 1.204$$

The *IRR* can be interpreted as the rate of change in the predicted number of outcome events for a 1-unit change in the predictor. In general,

- if $\hat{\beta} = 0$, then $\widehat{IRR} = 1$, and this indicates that there is no relationship between the predictor and the outcome (controlling for other predictors that may be included in the model);
- if $\hat{\beta} > 0$, then $\widehat{IRR} > 1$, and this indicates a positive relationship between the predictor and the outcome (controlling for any other predictors in the model), and
- if $\hat{\beta} < 0$, then $\widehat{IRR} < 1$, which indicates a negative relationship between the predictor and the outcome (controlling for any other predictors in the model).

*IRR*s for the estimates from Model 5.1 on the Bogotá data are shown in Table 5.4. The *IRR* for *Vendors1000* is about 1.2, which means that the number of police operations is predicted to increase by a factor of 1.2 for every additional 1000 street vendors in a Bogotá district (keeping population and proportion of lower class residents constant). We may express the same thing by saying 'our model predicts that the number of police operations will be 20% larger in a district that has 1000 more street vendors than another district of equal size and with the same proportion of lower class residents'.

To give another example: The *IRR* 0.986 for *Lower10* means that the number of police operations is predicted to be smaller by a factor of 0.986 (or to be 1.4% smaller) in a district whose proportion of lower class residents is 10 percentage points higher than in another district with the same population size and number of street vendors.

Let's also note the confidence intervals (we will look at them formally later, in the section 'Model Comparison and Inference'). The 95% CI for the *IRR* of *Vendors1000* is about 1.15 to 1.26. This indicates that a difference of 1000 street vendors between two districts is likely to be associated with a 15% to 26% difference in the number of police operations, keeping population size and percentage of lower class residents constant. So we can be reasonably confident in concluding that more street vendors are associated with more police operations in Bogotá. On the other hand, the 95% CI for the *IRR* of *Lower10* is about 0.94 to 1.03. This indicates that our results are consistent with a negative association, a positive association, or no association between the number of police operations and *Lower10*. So we can't be confident from these results that the poverty of a district is related to the number of police operations. Of course, this in itself does *not* mean that we have proved the absence of a relationship. However, overall the results do seem to confirm Holland's hypothesis, which posited that in centralised Bogotá, the number of police operations in a district should be related to the number of street vendors, but not to the poverty level.

Visualising the estimated regression line from a Poisson model

We might visualise the prediction from a Poisson model as in Figure 5.5. In this example, the regression line is calculated for a district with average population and average percentage of lower class residents. This is a choice made by the researcher (myself, in this case) to illustrate the main relationship of interest, which in this case is the relationship between the number of street vendors and the number of police operations.

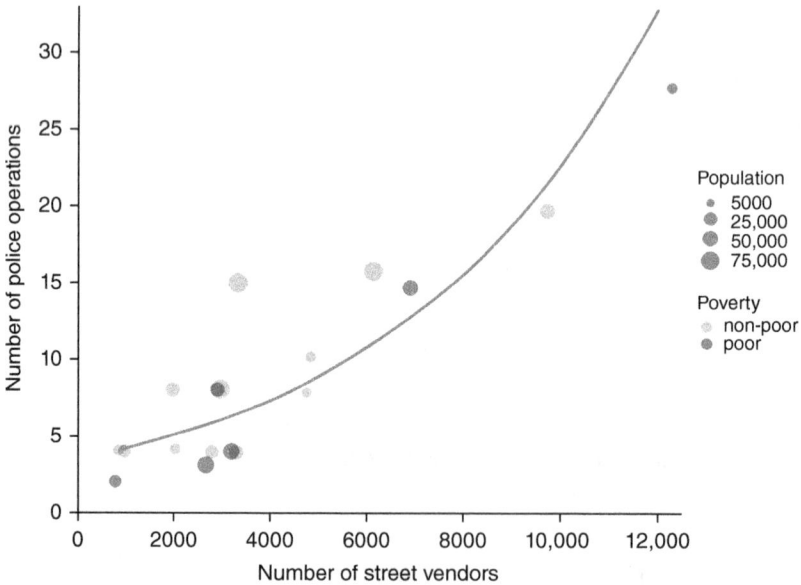

Figure 5.5 The number of police operations in 19 districts in Bogotá, by number of street vendors, population and poverty – and predictions for average district size and average poverty level from Model 5.1

Note. The line shows predicted numbers of police operations are given for districts with average population size (35,970) and average percentage of lower class residents (39.6%) in Bogotá. Dots were jittered out of position slightly to make visible multiple districts with the same number of vendors and police operations.

More generally, the line illustrates the shape of a positive relationship between a predictor and an outcome in a Poisson regression: as the number of street vendors goes up, the number of police operations is predicted to rise exponentially. This shape is a consequence of the log transformation of the outcome variable.

Negative binomial regression

When the assumption of equidispersion is not realistic – that is, if our outcome is *overdispersed* – a Poisson model may not be adequate. Instead of Poisson regression, we may propose a negative binomial regression model. One version of the negative binomial regression model is

$$\log(\mu_i) = \beta_0 + \beta_1 X_{1i} + \beta_2 X_{2i} + \ldots + \beta_k X_{ki}$$

$$Y_i \sim NegBin(\mu_i, \alpha)$$

$$\text{var}(Y_i) = \mu_i + \alpha \mu_i^2$$

This negative binomial model looks very similar to a Poisson model. Again we are using a logarithmic transformation of the outcome. The difference is that here we are expecting the observed numbers of events to follow a negative binomial (NegBin) distribution, rather than a Poisson distribution. As we saw in the section 'The Negative Binomial Distribution', the negative binomial distribution has two parameters, μ and α. Usually, only the mean μ is expected to be related to the predictors, but the dispersion parameter α is also estimated.[vi] The model must specify how we expect α to relate to the variance. Here I choose to specify that $\text{var}(Y_i) = \mu_i + \alpha \mu_i^2$. This is called the NB2-parameterisation. Other options exist; Box 5.2 provides more information. At any rate, in negative binomial regression the predicted variance var(Y_i) depends on both α and the predicted mean. This implies that we expect observations around the predicted mean to be heteroscedastic, rather than homoscedastic as in linear regression.

The coefficients from a negative binomial regression, as presented here, are interpreted in the same way as those from a Poisson regression. Exponentiating the coefficients gives us an incidence rate ratio, which we can interpret in the same way in negative binomial regression as we did in Poisson regression. Finally, I should note that there are different versions of negative binomial regression, which differ in the transformation used for the outcome and in the way the relation between the variance, the mean and the dispersion parameter is specified.

[vi]This is similar to linear regression, which not only relates the mean to a set of predictors but also estimates the error variance σ^2 (or, equivalently, the residual standard error σ).

In this book, I will only present one version in detail, but Box 5.2 gives some information on other options.

As an example, let's use negative binomial regression to investigate predictors of the number of police operations in districts in Lima.[vii] The Lima data are illustrated in Figure 5.6. The picture here is rather different to the one we saw in Bogotá (see Figure 5.4). In Lima, the number of police operations is clearly related to district poverty: there are fewer police operations in poor districts than in non-poor districts, even though the poor districts tend to have more street vendors. This, again, is consistent with Holland's theory: in decentralised Lima, district mayors would be expected to encourage police operations in richer districts, but not in poorer districts. The large dot on the right in Figure 5.6 suggests that there may be one outlier district, which is poor and has both a large number of street vendors and a large number of police operations.

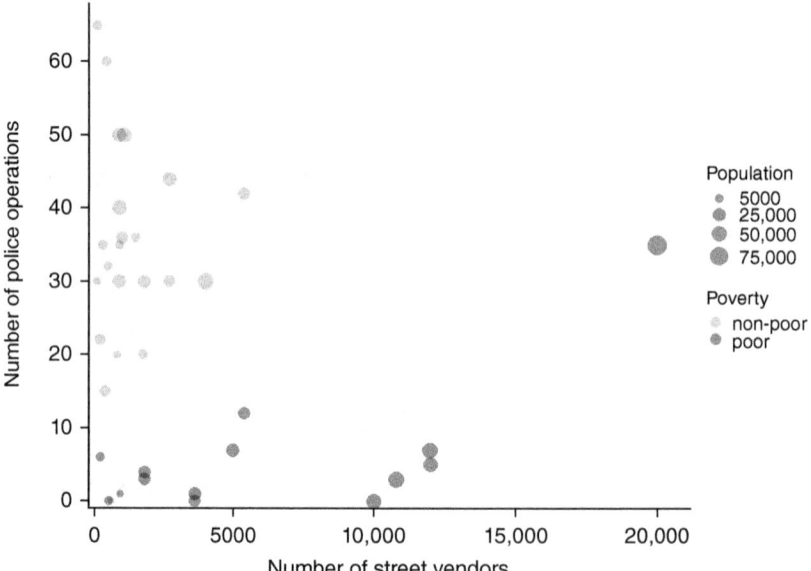

Figure 5.6 Number of police operations by number of street vendors in 36 Lima districts

Note. Each dot represents one city district. Data provided by Alisha C. Holland via the Harvard Dataverse (Holland, 2014). 'Poor' districts are defined to be those with 50% or more lower class residents (cf. Holland, 2015).

[vii]We will see later, in the section 'Model Comparison and Inference', why I have chosen negative binomial regression for the Lima data and Poisson regression for Bogotá.

Box 5.2

Negative Binomial Regression: Some More Options

This book presents the most popular version of negative binomial regression. But the specification of a negative binomial model is quite flexible. In particular, we can vary the specification in two types of ways:

Different link function. The negative binomial model as defined in the section 'Negative Binomial Regression' uses a logarithmic transformation of the outcome, $\log(\mu_i)$. A transformation of an outcome is also called a *link function*, or *link*: the idea is that we link the outcome to a set of predictors via a transforming function. The negative binomial model I present in this book is based on the log link. Other link functions might be used. For example, the link function $\log\left(\frac{\alpha\mu_i}{1+\alpha\mu_i}\right)$ is sometimes used, as is the so-called *identity link* (which involves no transformation). For a particular data set, a particular link function might yield a better fit to the data than others.

Different variance parameterisation. The model I chose to present specifies the variance to be $\mathrm{var}(Y_i) = \mu_i + \alpha\mu_i^2$. This is called the NB2-parameterisation. An alternative is the NB1-parameterisation, which instead specifies $\mathrm{var}(Y) = \mu_i + \alpha\mu_i$. Other variance functions are also possible. Different parameterisations might achieve a better fit to a particular data set.

See Hilbe (2012, 2014) for a more thorough discussion of these issues.

To formally investigate the relationships between police operations and the predictors in Lima, we specify a negative binomial regression model.

Model 5.2:

$$\log(\mu_i) = \beta_0 + \beta_1 Vendors1000_i + \beta_2 Population1000_i + \beta_3 Lower10_i$$

$$Operations_i \sim NegBin(\mu_i, \ \alpha); \ \mathrm{var}(Operations_i) = \mu_i + \alpha\mu_i^2$$

The variables *Operations, Vendors1000, Population1000* and *Lower10* are defined as for Model 5.1.

The results of estimating Model 5.2 on the Lima data are presented in Table 5.5. The coefficients from a negative binomial regression are interpreted in the same way as those from a Poisson regression. Once again, our outcome is log transformed, so the raw coefficients do not have an intuitive interpretation. However, we can transform them into *IRRs* to make them more straightforward to understand.

Table 5.5 Estimates from a negative binomial regression of the number of police operations in 36 Lima districts on number of street vendors, district population and percentage of lower class residents (Model 5.2)

	Coefficient	SE	IRR	95% CI
Intercept	3.977	0.309		
Vendors (per 1000)	−0.037	0.056	0.964	[0.864, 1.075]
Population (per 1000)	0.032	0.012	1.032	[1.008, 1.056]
Percentage lower class (per 10%)	−0.443	0.075	0.642	[0.554, 0.744]
Dispersion (α)	0.560			

Note. Confidence intervals were calculated via normal approximation. *SE* = standard error; *IRR* = incidence rate ratio; CI = confidence interval.

Table 5.5 also gives the estimate for the dispersion parameter α, which is $\hat{\alpha} = 0.56$. This tells us that the estimated variance of the counts around the predicted mean is $\widehat{var(Y_i)} = \mu_i + 0.56\mu_i^2$. Generally, the larger α, the larger the variance relative to the mean. In most applications, the dispersion parameter itself does not help us to answer our research questions. So we may not be interested in its estimate. However, it is important to include it in the model if a negative binomial distribution fits the data better than a Poisson distribution. (Model selection is discussed below, in the section 'Model Comparison and Inference'.)

Let's return our attention to the effects of the predictors. The results for Lima are quite different to those we obtained for Bogotá from Model 5.1. The percentage of lower class residents in Lima is negatively related to the number of police operations. The *IRR* is 0.642, with a 95% CI [0.554, 0.744]. This means that comparing two districts with equal population sizes and the same number of street vendors, which are 10 percentage points apart in terms of lower class residents, the poorer district is predicted to have 35.8%[viii] fewer police operations than the richer district (95% CI: between 44.6% and 25.6% fewer police operations). So in contrast to Bogotá, in Lima there is good evidence that fewer police operations are conducted in poorer districts.

Also in contrast to Bogotá, in Lima the number of police operations is negatively related to the number of vendors: the more vendors there are in a district, the fewer police operations are made against them, controlling for population size and the proportion of lower class residents. However, note that the 95% CI for the *IRR* is [0.864, 1.075], so the data are consistent with a mild positive as well as a negative association, and indeed with no association at all.

[viii]The *IRR* for *Lower10* is estimated to be 0.642. To calculate the percentage reduction, we note that 1−0.642 is 0.358, or 35.8%.

We can visualise the predictions from a negative binomial regression in the same way as for Poisson regression. Figure 5.7 gives an example. The line illustrates the predicted number of police operations for a district with average population and average number of street vendors, dependent on the percentage of lower class residents.

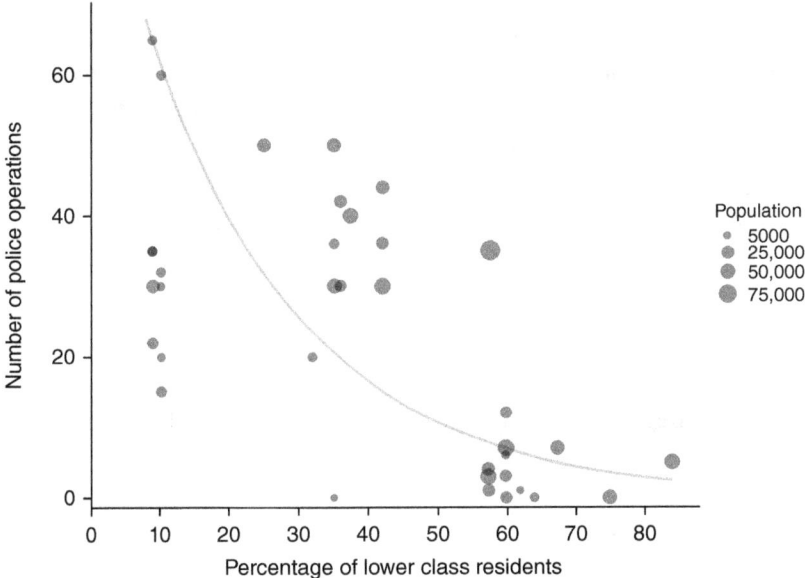

Figure 5.7 The number of police operations in 36 districts in Lima, by percentage of lower class residents and population – and predictions for districts of average size and with average number of vendors from Model 5.2

Note. The line shows predicted numbers of police operations are given for districts with average population size (22,626) and average number of street vendors (3238) in Lima. Larger dots tend to be above the line and smaller dots below the line – this indicates the influence of population size on the number of police operations: larger districts tend to have more, so the line for an average size district underestimates the number of police operations there.

More generally, the line in Figure 5.7 also illustrates the shape of a negative relationship implied by the log transformation of the outcome: as the percentage of lower class residents increases, the predicted number of police operations declines ever more slowly towards zero (but never quite reaches zero).

We have now defined Poisson and negative binomial regression, and seen an applied example of each. In the section 'Model Comparison and Inference', we will look at methods for deciding which of the two models may be more appropriate for a given data set. Before we do so, however, we must look at some specific situations that may arise in count variables, for which we may need to modify our models. In particular, we shall look at two types of situations:

- *zero-truncation*, where our research design prevents us from observing any zero counts, and
- *excess zeroes*, where our count variable has more zero counts than predicted by either the Poisson or negative binomial distribution.

Zero-truncation: when no zeroes are observed

There are some situations in which we may not be able to observe zero events. For example, imagine that our outcome is the *number of appointments attended by clients seeing a psychotherapist*. We may only have data on clients who have come to at least one appointment. There may also exist *potential* clients, who may have made a first appointment but never attended one. Now imagine that these 'zero appointments clients' did not consent to have their data stored, so we don't know how many of them there are. Thus, the minimum number of appointments in our data set is 1. Zeroes may well exist, but they are not observed in our study.

A count variable whose zeroes are unobserved is called *zero-truncated*. Zero-truncation can occur in any count variable, including those following a Poisson and those following a negative binomial distribution. Figure 5.8 shows an example of a hypothetical zero-truncated negative binomial distribution.

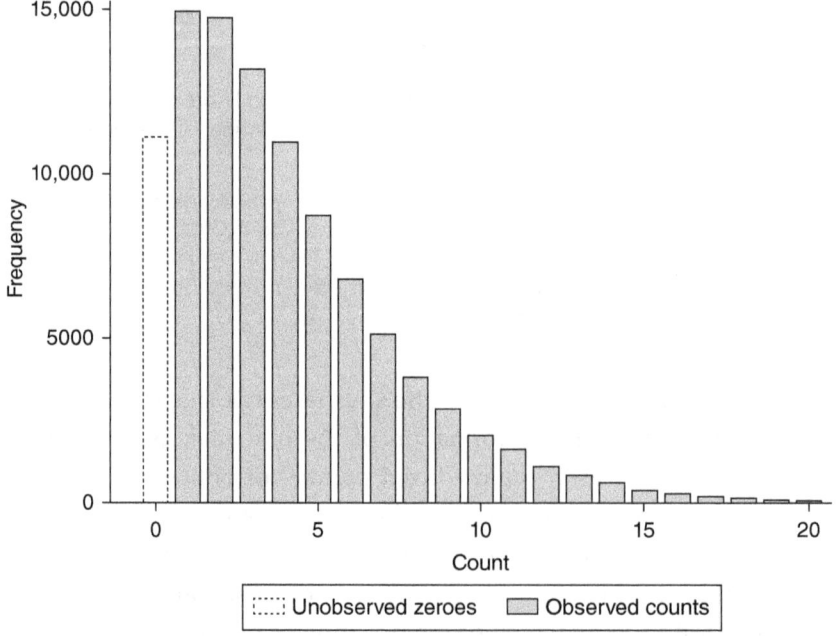

Figure 5.8 Zero-truncated negative binomial distribution

Note. Hypothetical data. This graph shows a negative binomial distribution with mean $\mu = 4.0$ and variance $\sigma^2 = 12.1$, including the zeroes. If the zeroes are not contained in your data set, you only see the zero-truncated data shown by the solid grey bars. Your observed mean would be $\bar{x} = 4.5$, and your observed variance $s^2 = 11.4$.

Absence of zeroes in a given data set is not in itself a reason to assume a zero-truncated distribution. In count distributions with large means, the probability of observing zero events may be quite small. For example, in a Poisson distribution with a mean of 4.61 or higher, the probability of observing a zero is below 1%. Thus, in a small sample it is possible that we do not observe any zeroes, even if no zero-truncation is involved in the data generation. On the other hand, if you know or suspect that your research design or data generation process *systematically* prevented zeroes from being observed, then you should consider a zero-truncated model. An application of zero-truncated negative binomial regression can be found in an analysis of hospital admissions for ischaemic stroke (Lee et al., 2003). The outcome variable is the number of hospital admissions, but only patients with at least one hospital admission are included in the sample.

If our outcome variable is a zero-truncated count, we may use either a zero-truncated Poisson regression or a zero-truncated negative binomial regression model. In either of these two models, predicted values will have a minimum value of 1. Otherwise the interpretation of coefficients is the same as in ordinary Poisson or negative binomial regression. So I won't give an example analysis here. As you will see below, zero-truncated count models are also used as part of another type of model called a hurdle model, which we will discuss in the next section. So you will see an example of a zero-truncated model there, albeit as part of a larger and more complex model.

Too many zeroes: zero-inflation and hurdle models

Sometimes we encounter situations where the count outcome we wish to model seems to follow a Poisson or a negative binomial distribution, except that the number of zeroes is much larger than expected under either of those models. In such a situation, statisticians say that the count variable has **excess zeroes**. For example, your distribution may look as illustrated in Figure 5.9. The grey bars show the (hypothetical) observed data. The observed mean count is 1.516. If we plot a Poisson distribution with mean 1.516, we see that the distribution does not fit the data well. There are many more observed zeroes than predicted by the Poisson distribution.

Sometimes, a Poisson or negative binomial regression model can fit such data well: this is the case if one of our covariates identifies a group of cases with a very low mean count, and if many of the zeroes come from this group. Then, although the outcome variable taken on its own seems to have excess zeroes, once the effects of the predictors have been taken into account, the distribution of counts around the predicted means follows a Poisson or negative binomial distribution well enough. However,

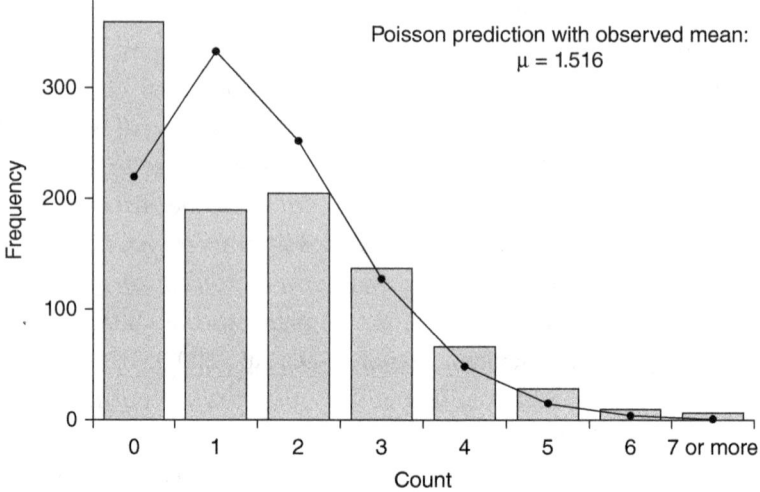

Figure 5.9 Example of a count distribution with excess zeroes

Note. Hypothetical data. The grey bars show how many times there were 0, 1, 2, . . . counts. The black dots show the predicted proportions under the assumption that the data come from a Poisson distribution with the observed mean (1.516). The data and the predictions do not match. There are many more zeroes in the data than predicted by the Poisson distribution.

there are many cases when a Poisson or negative binomial regression does not fit data with excess zeroes well.

So we need to consider models that are specifically designed for count data with excess zeroes. Two types of models are often used for this purpose: zero-inflated models and hurdle models. They both attempt to account for the excess zeroes in different ways. Both can be applied to either Poisson or negative binomial distributions. We will discuss zero-inflated distributions first.

Zero-inflated count distributions

A zero-inflated count distribution comes about when zero counts arise in two different ways. Consider, as an example, a study of cannabis use. Suppose that the researchers ask their respondents how many joints they have smoked in the last week. Now, many people never smoke cannabis, so their answer is always zero, no matter in which week we happen to ask them. On the other hand, some portion of our population do smoke cannabis, with varying frequency. Their answer to the question 'how many joints have you smoked last week?' might be 0, 1, 2, and so forth. Importantly, a cannabis user can truthfully give the answer '0', depending on which week we happen to ask about. Thus, the zeroes in our data come from two sources: (1) zeroes from non-cannabis users and (2) zeroes from cannabis users who did not smoke a joint in the reference week.

We call these two kinds of zeroes 'structural zeroes' and 'sampling zeroes'. Structural zeroes are observed on cases that always have zeroes (e.g. non-users of cannabis), while sampling zeroes are observed on cases that have a non-zero mean, but may happen to have an observed count of zero on a particular occasion (e.g. an occasional cannabis user whom we happened to ask about a week in which they did not have a joint).[ix]

Figure 5.10 illustrates the idea of a zero-inflated distribution. The light grey bars represent a Poisson distribution, the dark grey bar adds structural zeroes to the sampling zeroes from the Poisson.

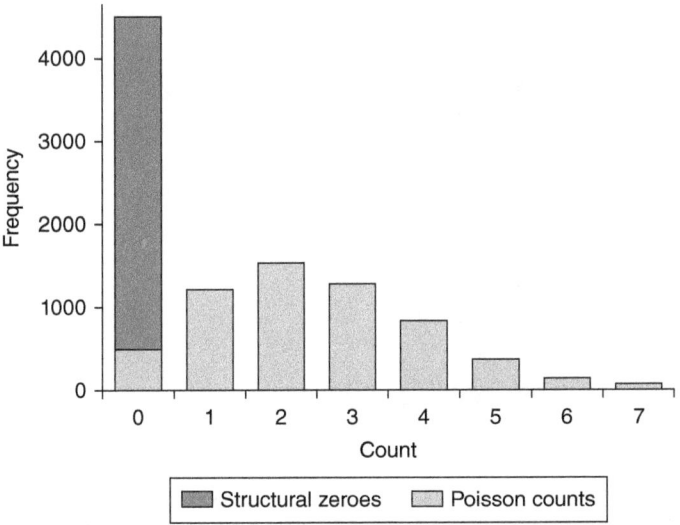

Figure 5.10 A zero-inflated Poisson distribution (assumed by zero-inflated Poisson regression)

Count distributions with hurdles

There is a second way in which excess zeroes may come about. For example, imagine a doctor who sometimes refers her mentally unwell patients to a psychotherapist. We may wish to study the number of appointments with a psychotherapist that the referred patients attend. Now, some patients, despite the doctor's recommendation,

---

[ix]One way to understand why these zeroes are called sampling zeroes: consider that we sample 1 week in the life of a cannabis smoker. Let's imagine an occasional smoker of cannabis, who smokes 0.7 joints per week on average, and let's assume that the number of joints our smoker has per week follows a Poisson distribution. Then it follows (cf. Box 5.1) that in 49.66% of weeks, they smoke no joints. Thus, in our study, we have about a 50% chance to sample a zero from this cannabis smoker's life.

never go to see a therapist; they attend 0 appointments. For those patients who do go and see a therapist, however, the number of appointments must be at least 1.

We might say that the distribution of the number of appointments is subject to a *hurdle*: if the hurdle is not cleared, if the patient never goes to see a therapist, the number of events is 0. Once the hurdle is cleared, the number of events must be at least 1. So in a hurdle model, all zeroes are structural zeroes; there are no sampling zeroes. A count distribution resulting from a hurdle model is illustrated in Figure 5.11. The light grey bars represent a zero-truncated Poisson distribution. The dark grey bar represents the structural zeroes. The proportion of structural zeroes is independent of the zero-truncated Poisson distribution.

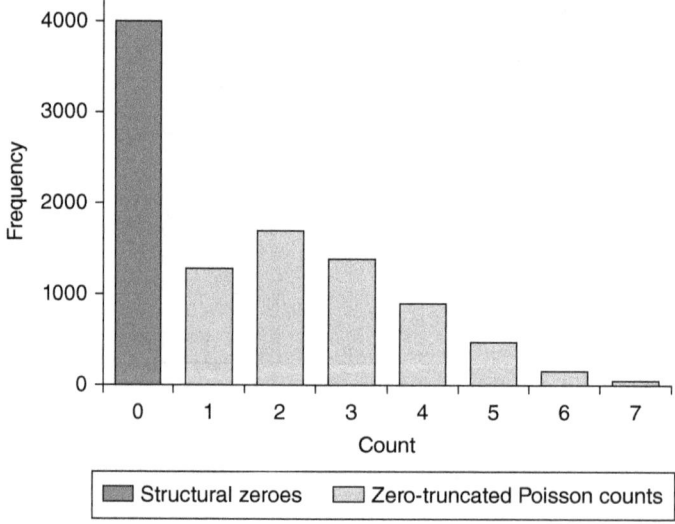

Figure 5.11 A mixture distribution of zeroes and a zero-truncated Poisson (assumed by a Poisson hurdle model)

Models for outcomes with excess zeroes

Modelling an outcome with excess zeroes is a little more complicated than an ordinary count model. Such models consist of two parts:

- One part predicts structural zeroes.
- The other part predicts the remaining counts.

This section will introduce zero-inflated and hurdle models for both Poisson and negative binomial count distributions. As we will see, in both zero-inflated and hurdle models, we use logistic regression to predict structural zeroes. But the zero-inflated and hurdle models differ in how we model the counts:

- In a zero-inflated model, the counts are modelled with a Poisson or negative binomial distribution, and predicted counts have a theoretical minimum of zero (because there can be sampling zeroes as well as structural zeroes).
- In a hurdle model, the counts are modelled with a zero-truncated Poisson or zero-truncated negative binomial distribution, and predicted counts have a theoretical minimum of 1 (because there can be no sampling zeroes).

As an example, we will consider once again Holland's study on police operations, this time using the data from Santiago de Chile. Santiago is a decentralised city, so we would expect the results to be more similar to Lima than to Bogotá.

Figure 5.12 shows data from 34 districts in Santiago. The mean observed number of police operations per district in Santiago is 2.7, and the variance is 24.4. Since the variance is much larger than the mean, we might wish to consider a negative binomial model for these data. There is also evidence of excess zeroes. Eighteen districts had 0 police operations; the remaining 16 districts had between 1 and 16 police operations.[x] So more than half of our outcome values are zeroes. This is many more than predicted by a negative binomial distribution with the observed mean and variance. So a zero-inflated or hurdle model should be considered. We will discuss each model in turn. In the examples below, I will first show zero-inflated Poisson and hurdle Poisson models, because their mathematics are simpler to explain than those of their negative binomial counterparts. However, the logic and principles of interpretation are essentially the same for the Poisson and negative binomial versions of models with excess zeroes.

Zero-inflated models

The zero-inflated count model works as follows. We assume that the observed number of events is the result of two distributions:

- One distribution determines whether we observe a structural zero or not. We use π (the Greek letter *pi*) to denote the probability of observing a structural zero. We model this probability with a logistic regression.
- For cases where we don't have a structural zero, we model the probability of observing 0, 1, 2, . . . events via a count model (either Poisson or negative binomial).

---

[x]It is difficult to see all the zeroes in the plot, since many districts have very similar numbers of vendors. To make individual districts easier to see, I have jittered the points a bit horizontally.

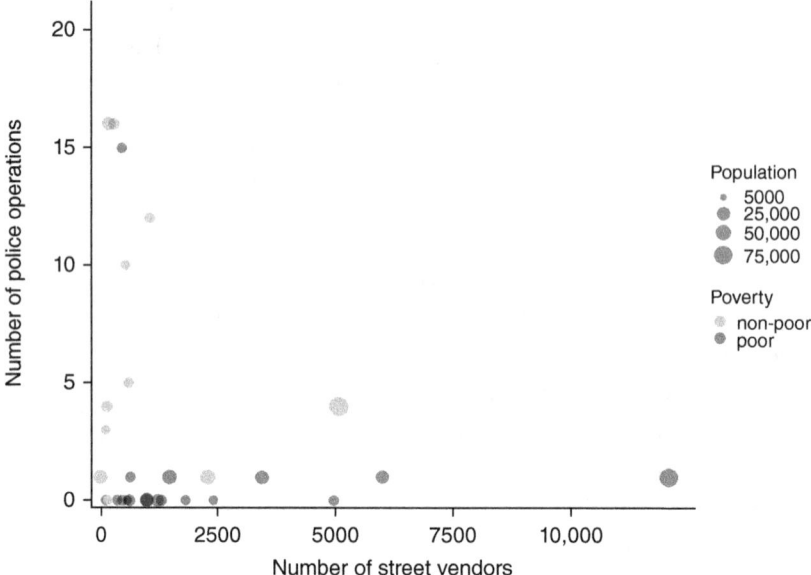

Figure 5.12 Number of police operations by number of street vendors in 34 districts in Santiago

Note. Data provided by Alisha C. Holland via the Harvard Dataverse (Holland, 2014). 'Poor' districts are defined to be those with 50% or more lower class residents (cf. Holland, 2015). Points have been jittered horizontally to make overlapping districts easier to see.

Mathematically, a zero-inflated Poisson model can be expressed as follows:

$$P(Y_i = 0) = \pi_i + (1 - \pi_i)e^{-\mu_i}$$

$$P(Y_i = k) = (1 - \pi_i)\frac{\mu_i^k e^{-\mu_i}}{k!}, \; k \geq 1$$

The first equation describes the probability of observing zero events. This probability is the sum of the probability of a structural zero (π_i) and the probability of a sampling zero $[(1-\pi_i)e^{-\mu_i}]$. The second equation describes the probability of observing 1, 2, 3, . . . or more events. (The zero-inflated negative binomial model looks very similar, but the equations have more complicated right-hand sides.) In this model, there are two parameters, π and μ, each of which appears in both equations. But we model these parameters separately.

The probability π of having a structural zero is modelled via a logistic regression:

$$\text{logit}(\pi_i) = \gamma_0 + \gamma_1 X_{1i} + \gamma_2 X_{2i} + \dots$$

The mean μ of the counts that are not structural zeroes is modelled via a Poisson regression:

$$\log(\mu_i) = \beta_0 + \beta_1 X_{1i} + \beta_2 X_{2i} + \dots$$

A few notes on these two models and the way I have written them:

- I have called the coefficients in the logistic part $\gamma_0, \gamma_1, \gamma_2, \dots$ to make clear that these are not the same coefficients as those in the Poisson part of the model.
- The predictor variables do not need to be the same in the two model parts. In the equations above, X_1 and X_2 appear as predictors in both the logistic and the Poisson parts, but I could have chosen to have entirely different predictors in each part of the model, or to use some predictors in both model parts, and other predictors in only one of them.

If this all sounds a bit difficult, an example might help. Let's define a zero-inflated Poisson model of the Santiago police operations data: Let π_i be the probability of structural zero police operations in a Santiago district, and let μ_i be the mean number of police operations that are not structural zeroes. Then, for example, I might specify the following model:

Model 5.3:

Logistic part:

$$\text{logit}(\pi_i) = \gamma_0 + \gamma_1 Vendors1000_i + \gamma_2 Population1000_i + \gamma_3 Lower10_i$$

Poisson part:

$$\log(\mu_i) = \beta_0 + \beta_1 Lower10_i$$

The predictor variables are defined as in Models 5.1 and 5.2. This model reflects the following hypotheses:

- *Logistic part:* the probability of having structural zero police operations depends on the number of vendors in a district, its population size, and its poverty level.
- *Poisson part:* for the districts that are not structural zeroes, the number of police operations depends only on the district's poverty level.

If we estimate this model on the Santiago data, we obtain the estimates shown in Table 5.6.

Table 5.6 shows the Poisson part and the logistic (zero-inflation) part of the model separately. I have also transformed the coefficients to give a rate ratio for the Poisson part and odds ratios for the zero-inflation part of the model.

Table 5.6 Estimates from a zero-inflated Poisson regression of the number of police operations per district in Santiago

Poisson	Coefficient	SE	IRR	95% CI
Intercept (β_0)	2.141	0.193		
Lower (per 10%) (β_1)	−0.122	0.054	0.89	[0.80, 0.98]
Logistic (Zero-Inflation)	**Coefficient**	**SE**	**OR**	**95% CI**
Intercept (γ_0)	−1.313	2.017		
Vendors (per 1000) (γ_1)	−0.015	0.478	0.99	[0.39, 2.51]
Population (per 1000) (γ_2)	−0.273	0.123	0.76	[0.60, 0.97]
Lower (per 10%) (γ_3)	1.144	0.444	3.14	[1.31, 7.50]

Note. SE = standard error; IRR = incidence rate ratio; OR = odds ratio; CI = confidence interval.

The coefficients in both model parts are estimated in a single procedure. This implies that each coefficient estimate is conditional on how both parts of the model were specified. Thus, for example, our interpretation of the *IRR* for *Lower10* should be that it is the estimated effect of *Lower10* on the number of police operations controlling for structural zeroes, where the structural zeroes are being predicted by *Lower10*, *Vendors1000* and *Population1000*.

Another issue we need to pay attention to when interpreting the slope coefficients, *IRRs*, and odds ratios is the direction of the effect in each model part. The zero-inflation part is a logistic regression predicting the zeroes, so a positive coefficient means that a predictor makes zeroes more likely, while a negative coefficient means that a predictor makes zeroes less likely. Thus the direction in which the coefficients should be interpreted differs between the two model parts. In the Poisson part, a positive coefficient (or *IRR* > 1) means 'the larger *X*, the larger the predicted number of events'. In the logistic part, a positive coefficient (or *OR* > 1) means that 'the larger *X*, the smaller the predicted number of events'.[xi] To think this through for the logistic part of the model, let's look at two examples:

- The positive coefficient (1.144) of *Lower10* in the logistic part means that the poorer the district, the *higher* the predicted log odds of *zero* police operations. The odds ratio is 3.14, so from this model we estimate that a 10 percentage point difference in the proportion of lower class residents is associated with 3.14 times higher odds of zero police operations. So poorer districts are more likely to have structural zero police operations than richer districts, controlling for all other parameters in the model.

[xi]I recognize that this may sound confusing. I would have loved to define the logistic part of the zero-inflated model differently, so that a positive coefficient always means 'more predicted police operations'. But if I had done that, I would have been inconsistent with all other books on this topic that I know, and also with the output from all statistical software that I know. That would have led to more confusion once readers of this book consult other sources or use statistical software to apply zero-inflated models.

- The negative coefficient (–0.273) for *Population1000* means that the more populous a district is, the lower are the predicted log odds of zero police operations. The odds ratio is 0.76, so from this model we estimate that a 1000-people difference in population size is associated with a 24% reduction in the probability of zero police operations. So more populous districts are less likely to have structural zero police operations than less populous districts, controlling for all other parameters in the model.

In the Poisson part, we can interpret the coefficients and the *IRR*s as usual for a Poisson regression: in our case, the estimated *IRR* for *Lower10* is 0.89, so the model estimates that a 10 percentage point difference in the proportion of lower class residents is associated with an 11% reduction in the mean police operation count, after the structural zeroes have been accounted for. So the model estimates that poorer districts, if they have any police operations at all, tend to have fewer police operations than richer districts.

Overall, from this model it looks once again as though Holland's hypothesis is borne out by the data. Santiago is a politically decentralised city, so Holland's theory predicts that the number of police operations depends more on the poverty level of the district than on the number of vendors. This is what we find: both in the zero-inflated and the Poisson parts, there is evidence that there are fewer police operations in districts with a higher percentage of lower class residents. On the other hand, there was little evidence from this model that the number of vendors is related to the number of police operations in Santiago, since the odds ratio for vendors in the zero-inflation part is estimated to be close to 1, and since the confidence interval includes the value 1. Of course, in order to test the hypothesis thoroughly, we should examine whether the number of vendors makes a difference to the prediction in the Poisson part of the model also. But I won't explore that here, to keep the example simple.

Hurdle models

Now we turn to hurdle models, the other method of accounting for excess zeroes in a count regression. In a hurdle model, we assume that the number of events we observe is the result of two distributions: one distribution determines whether we observe a zero or a non-zero number of events, the other determines what the non-zero counts are (1, 2, 3, . . .).

We use π (the Greek letter *pi*) to denote the probability of having a *non-zero* count.[xii] We model this probability with a logistic regression. Then, for cases that are not zero (cases that 'clear the hurdle'), we model the probability of observing counts 1, 2, 3, . . .

[xii]So in a hurdle model, π is the probability of the number of events being *not* zero, in contrast to the zero-inflated model, where π denotes the probability of a structural zero.

via a zero-truncated count model (either zero-truncated Poisson or zero-truncated negative binomial). Mathematically, the Poisson hurdle distribution can be written like this:

$$P(Y_i = 0) = 1 - \pi_i$$

$$P(Y_i = k) = \frac{\pi_i \mu_i^k}{k!(e^{\mu_i} - 1)}, \ k \geq 1$$

The first equation describes the probability of structural zeroes. The second equation describes the probabilities of the non-zero counts. Two parameters appear in these equations, π and μ. The parameter π features in both equations, but μ features only in the second equation.[xiii] (The negative binomial hurdle distribution looks similar to the Poisson hurdle distribution, but the second equation has a more complicated right-hand side.)

As in zero-inflated regression, we model the parameters separately. The probability π of clearing the hurdle is modelled via a logistic regression:

$$\text{logit}(\pi_i) = \gamma_0 + \gamma_1 X_{1i} + \gamma_2 X_{2i} + \dots$$

The parameter μ for the non-zero counts is modelled via a zero-truncated count regression:

$$\log(\mu_i) = \beta_0 + \beta_1 x_{1i} + \beta_2 x_{2i} + \dots$$

As in zero-inflated regression, we may use either the same or different sets of predictors in the two model parts.

For example, I might define a model of the Santiago police operations data as follows: Let π_i be the probability of non-zero police operations in a Santiago district, and let μ_i be the parameter in a zero-truncated Poisson distribution of police operations in districts that do not have zero counts. Then, for example, I might specify the model:

Model 5.4:

Logistic part:

$$\text{logit}(\pi_i) = \gamma_0 + \gamma_1 Vendors1000_i + \gamma_2 Population1000_i + \gamma_3 Lower10_i$$

[xiii]Because the second equation relates to a zero-truncated model, the parameter μ is not the mean of the non-zero counts. Instead, the mean of the non-zero counts is . So the mean of the non-zero counts is larger than the parameter μ.

Zero-truncated Poisson part:

$$\log(\mu_i) = \beta_0 + \beta_1 Lower10_i$$

Estimates from fitting this model on the Santiago data are shown in Table 5.7. The logistic (hurdle) part of the model predicts the probability of having a *non*-zero count. Thus, in contrast to the zero-inflated model (compare Table 5.6), the coefficients from both parts can be interpreted in the same direction. A positive coefficient means that a predictor variable is associated with a larger number of events: a larger probability of being non-zero, or a larger predicted number of events among the non-zeroes, respectively. A negative coefficient means the opposite. Notice, therefore, that the coefficients in the hurdle part of the model in Table 5.7 have opposite signs to their counterparts in the zero-inflated part of Table 5.6. Other than that, the results are very similar.

Table 5.7 Estimates from a Poisson hurdle model predicting police operations in 34 Santiago districts (Model 5.4)

Zero-Truncated Poisson	Coefficient	SE	IRR	95% CI
Intercept (β_0)	2.142	0.193		
Lower10 (β_1)	−0.122	0.054	0.88	[0.80, 0.98]
Logistic (Hurdle)	**Coefficient**	**SE**	**OR**	**95% CI**
Intercept (γ_0)	1.349	2.004		
Vendors (γ_1)	0.024	0.470	1.02	[0.41, 2.58]
Population (γ_2)	0.270	0.119	1.31	[1.04, 1.65]
Lower10 (γ_3)	−1.149	0.443	0.32	[0.13, 0.76]

Note. SE = standard error; *IRR* = incidence rate ratio; *OR* = odds ratio; *CI* = confidence interval.

A difference to the zero-inflated model is that in the hurdle model, both model parts are essentially separate. Because all zeroes are considered structural zeroes, the logistic regression part is independent of the zero-truncated count (e.g. Poisson) part of the model. Thus, in a hurdle model, if I change my predictors for one of the parts, this won't affect my coefficient estimates in the other part. Therefore, for example, the correct interpretation for the coefficient of *Lower10* in the zero-truncated Poisson part of Model 5.4 is that it is the effect of the percentage of lower class residents on the predicted non-zero counts, where the zero counts themselves are modelled entirely separately. This estimated effect of *Lower10* on the non-zero counts won't change depending on which set of predictors I choose to model the zeroes.

Final remarks on zero-inflated and hurdle models

Zero-inflated negative binomial regression and negative binomial hurdle regression work in the same way as their Poisson counterparts, except that the count part of the model assumes that the counts that aren't structural zeroes follow a negative binomial distribution instead of a Poisson distribution. Thus, in addition to the slope coefficients, in negative binomial hurdle and zero-inflated models, we also estimate the dispersion parameter. I won't show an example here, because the interpretation of the coefficients *IRRs* and *ORs* does not change substantially compared to the Poisson model described above. However, the next section ('Model Comparison and Inference') does feature an example of a negative binomial hurdle regression.

Whether a zero-inflated or hurdle model is more appropriate for your data may depend on several things. If you know the process by which excess zeroes are generated, then you may know which of the two models is a better representation of reality (see the examples in the Sections 'Zero-Inflated Count Distributions' and 'Count Distributions With Hurdles'). Sometimes we don't have a good theoretical justification for choosing one type of model over another, in which case we may use our data to investigate which model provides the better fit to our data. And that is the subject of the next section.

Model comparison and inference

So far in this chapter, we have mentioned eight types of models for count data. Table 5.8 gives an overview.

Table 5.8 Overview of types of models discussed in this chapter

	Zeroes as Expected	Excess Zeroes	No Zeroes
Equidispersed	Poisson	Zero-inflated Poisson *or* hurdle model with Poisson	Zero-truncated Poisson
Overdispersed	Negative binomial	Zero-inflated negative binomial *or* hurdle model with negative binomial	Zero-truncated negative binomial

The previous sections in this chapter should have given you an idea of the types of research situations in which each of these models is plausible, and sufficient information to interpret the coefficients. However, so far we have not dealt with the question of how to use our data to help us select the best model. That is what we are going to consider now.

There are two sorts of model comparisons to consider:

- Deciding between models assuming different outcome distributions, such as deciding between a Poisson and a negative binomial regression, or deciding whether a zero-inflation or a hurdle model should be used.
- Deciding between models featuring different sets of predictors. For example, within a negative binomial regression model, we might be interested in testing hypotheses about one or several predictor variables.

The two decisions are not independent of one another. The selection of predictors may influence which type of distribution appears to be plausible. For example, the distribution of an outcome may look overdispersed in a descriptive analysis, but once we have accounted for the influence of predictors, the residual variation may well fit a Poisson distribution. Similarly, we might observe more zeroes than expected by a negative binomial distribution in our outcome, but after accounting for the influence of predictors, we may find that the residual distribution fits a negative binomial model quite well, such that we do not need to employ a hurdle or zero-inflated model.

There is no single recipe for deciding on the best model, but the remainder of this section will suggest techniques that may help you choose between different plausible models by using the data. I would like to mention one issue at the outset. Previously in this book, we have used likelihood ratio tests to compare nested models with one another. As we will see, the likelihood ratio test (as defined in Chapter 2) will be appropriate for investigating nested models within the same model type. However, in general, likelihood ratio tests should not be used for comparing one of our eight model types with another. This is because

- models without zero-inflation are not nested within zero-inflated models,
- models without hurdles are not nested within hurdle models, and
- although the Poisson model is nested within the negative binomial model, the ordinary likelihood ratio test we used in Chapters 2 to 4 is not appropriate for comparing the two.

So we need to use other methods, and will turn to these methods next.

Investigating overdispersion: Poisson or negative binomial model?

Let's first consider the choice between a Poisson regression and a negative binomial regression (assuming, for now, that there is no zero-truncation and that there are no excess zeroes). This choice comes down to whether we think that our outcome variable is overdispersed or not. Several methods have been proposed for testing this. This section will only cover one method in detail, namely, the boundary likelihood ratio test. I will also mention some appropriate alternatives and caution against methods that are sometimes recommended but which, in fact, may systematically lead to wrong conclusions.

Boundary likelihood ratio test

First of all, let's establish that the Poisson model is nested within the negative binomial model. This can be seen to be true, because the Poisson distribution can be defined as a negative binomial distribution with dispersion parameter $\alpha = 0$ (see the section 'Distributions for Count Data'). Nonetheless, unfortunately we cannot use the ordinary likelihood ratio test to compare a Poisson to a negative binomial model (even though some authors erroneously advise this). This is because, for the negative binomial distribution, the dispersion parameter is non-negative by definition: $\alpha \geq 0$. Since the Poisson distribution results when $\alpha = 0$, the test value for the likelihood ratio test is the smallest possible value of α. Statisticians express this fact by saying that the test value for α is *on the boundary of the parameter space*. Under this condition, the *p*-value calculated by the ordinary likelihood ratio test is too large. This means that, using the ordinary likelihood ratio test, we would accept the null hypothesis (of equidispersion) too often in situations when it really should be rejected. Thus, we would be concluding too often that the Poisson model is adequate, in situations when the outcome is really overdispersed and a negative binomial model would be more appropriate. In other words, when testing a Poisson model (null hypothesis, $\alpha = 0$) against a negative binomial model (alternative hypothesis, $\alpha > 0$), the likelihood ratio test has an inflated type II error rate (for type II error, see *The SAGE Quantitative Research Kit*, Volume 3).

So we should not use the ordinary likelihood ratio test to investigate overdispersion. Instead we should use what has been called the *boundary likelihood ratio test* (Hilbe, 2014, p. 112). This is similar to an ordinary likelihood ratio test: in fact, the test statistic is calculated in the same way (as −2 times the difference between the log likelihoods of the two models being compared). However, the *p*-value is calculated differently. It turns out that the correct *p*-value for the boundary likelihood ratio test is half the *p*-value of the ordinary likelihood ratio test. Thus, pragmatically, we can use software to calculate the ordinary likelihood ratio test *p*-value, and divide the obtained number by 2 to get the boundary likelihood ratio test *p*-value.

Table 5.9 shows the results of two boundary likelihood ratio tests carried out to compare Poisson and negative binomial models for the Bogotá and Lima data, respectively. Note that for each model comparison, the predictors are the same (*Vendors1000*, *Population1000*, *Lower10*). The only difference between the two models in each comparison is whether a Poisson or negative binomial distribution is assumed for the outcome.

Table 5.9 Boundary likelihood ratio tests comparing Poisson with negative binomial models in Bogotá and Lima

		−2 × LL	df	Λ	df Diff.	Boundary LRT p-Value
Bogotá	Poisson (Model 5.1)	83.9234	15			
	Negative binomial	83.9233	14	0.0001	1	0.497
Lima	Poisson	467.28	32			
	Negative binomial (Model 5.2)	275.26	31	192.02	1	<0.0001

Note. The *p*-value for the boundary likelihood ratio test statistic can be calculated by halving the *p*-value obtained from an ordinary likelihood ratio test. For Bogotá, the test investigates the null hypothesis that the Poisson Model 5.1 predicts the number of police operations as well as a negative binomial model with the same predictors. For Lima, the test investigates the null hypothesis that a Poisson model with the same predictors predicts the number of police operations as well as the negative binomial Model 5.2. LL = log likelihood; *df* = degrees of freedom; Λ = likelihood ratio test statistic; Diff. = difference; LRT = likelihood ratio test.

Table 5.9 shows that, for the Bogotá data, there is little evidence that a negative binomial model would predict the outcome distribution better than the Poisson model ($p = 0.497$). For the Lima data, on the other hand, the test suggests that there is strong evidence that the negative binomial model gives better predictions than the Poisson data ($p < 0.0001$).

Note that the boundary likelihood ratio test is appropriate only for comparing Poisson and negative binomial models that have the same predictors. If you wanted to compare, say, a Poisson model with predictor X_1, and a negative binomial model with predictors X_1 and X_2, then you should not use the boundary likelihood ratio test, nor should you use any likelihood ratio test. Even though the two models would be nested, I don't know of a general way to derive the correct *p*-value for a case like this. I recommend instead the use of information criteria for such model comparisons. We shall discuss information criteria presently, in the section 'Information Criteria'.

Other methods for checking overdispersion

There are alternatives to the boundary likelihood ratio test for testing overdispersion. A thorough account is given by Hilbe (2014). Some software packages routinely display the *dispersion statistic* for a Poisson regression model. The dispersion statistic essentially compares the variance of the observed residuals with the variance implied by the Poisson model. In a well-fitted Poisson model, the dispersion statistic should be approximately equal to 1. Values larger than about 1.25 are taken as an indication of overdispersion (Hilbe, 2014), such that a negative binomial model ought to be considered. However, cut-offs such as 1.25 are always arbitrary, and I would recommend using this method only in conjunction with other methods, such as the boundary likelihood ratio test, information criteria (see the section 'Information Criteria') and graphical examination of the data and model residuals (see the section 'Model Evaluation').

Finally, some textbooks recommend a goodness-of-fit test for Poisson models. However, the use of this test is potentially problematic, and in general, I do not recommend it. Box 5.3 explains why.

Box 5.3

Goodness-of-Fit Test for Poisson Models: A Warning

One test that is often used to evaluate a Poisson model, but which in many situations is not appropriate, is the goodness-of-fit test. This tests the null hypothesis that a Poisson model fits the data well. Small p-values are taken to indicate that the Poisson model does not fit the data well. However, this test tends to be overly sensitive, particularly when the predicted means are relatively small. 'Overly sensitive' means that the test leads us to reject the null hypothesis too often in situations when the Poisson model is really appropriate (in other words, we have an inflated type I error rate – see *The SAGE Quantitative Research Kit*, Volume 3, on type I errors). So if we used the goodness-of-fit test to decide between a Poisson and a negative binomial model, we would end up choosing the negative binomial model too often unnecessarily. This issue has been discussed in statistics textbooks (Pawitan, 2013; Venables & Ripley, 2002), and an excellent blog post by Bartlett (2014) explains it in relatively simple terms. My advice is not to rely on the goodness-of-fit test, but to use the boundary likelihood ratio test instead to compare the Poisson model to a negative binomial model, or information criteria to compare the Poisson model to other plausible models, such as zero-inflated or hurdle models.

Information criteria

Information criteria are statistics that allow us to compare sets of either nested or non-nested models. They are often used in situations where statistical hypothesis tests are not available, but in principle there is nothing wrong with using information criteria in conjunction with, or even instead of, hypothesis tests, even in situations where a valid hypothesis test exists. In this book, I will introduce two information criteria: Akaike's Information Criterion (AIC) and the Bayesian Information Criterion (BIC). AIC and BIC have the same underlying idea: to provide the analyst with a measure of the quality of a model that balances goodness of fit with parsimony. Model fit is represented by the log likelihood; parsimony is represented by the number of parameters in the model.

Akaike's information criterion

Akaike's Information Criterion (AIC) is defined as

$$\text{AIC} = 2 \times k - 2 \times \text{LL}$$

where k is the number of parameters contained in the model, and LL is the log likelihood. You can think of the number of parameters as a penalty that a model pays for being larger. For example, a larger model may fit slightly better in a particular data set, but this may be outweighed by the model's complexity.

A smaller AIC indicates a better model. In general:

- If the larger of two models (i.e. the model with more parameters) does not improve the fit very much (has the same log likelihood as the smaller model, or only a slightly larger one), then the larger model will have the larger AIC, indicating that the smaller model should be preferred.
- If the larger model fits the data a lot better than the smaller one, then the larger model will tend to have the smaller AIC (because the increase in the log likelihood will outweigh the additional parameters). This would indicate that the larger model should be preferred.

Bayesian information criterion

The Bayesian Information Criterion is very similar to the AIC. The BIC is defined as

$$\text{BIC} = \log(n) \times k - 2 \times \text{LL}$$

where n is the sample size, k is the number of parameters and LL is the log likelihood. The idea behind the BIC is the same as that behind the AIC; and like with the AIC, a smaller BIC value indicates a better model. The difference is that in BIC, the penalty for additional parameters is larger, at $\log(n) \times k$ rather than $2 \times k$. This means that the BIC puts higher value on parsimony than the AIC. The BIC also recognises that the value of the log likelihood depends partly on the sample size, and so makes the size of the penalty term partly depend on the sample size also.

AIC and BIC in practice

As an example of how to use AIC and BIC for model comparison, consider the investigation into predictors of police operations against street vendors in Santiago. In the section 'Too Many Zeroes', we fitted zero-inflated and hurdle Poisson models to these data. How might we use the data to decide which model is more appropriate, or whether a plain Poisson or negative binomial model (without excess zeroes) is sufficient? Table 5.10 presents AIC and BIC values for six models estimated on the Santiago data.

Table 5.10 Comparison of six models predicting the number of police operations in Santiago

	Model	Log Likelihood	Parameters	AIC	BIC
Poisson	No excess zeroes	−233.64	4	475.28	481.39
	Zero-inflated (Model 5.3)	−75.97	6	163.94	173.09
	Hurdle (Model 5.4)	−75.93	6	163.86	173.02
Negative binomial	No excess zeroes	−137.63	5	285.26	292.89
	Zero-inflated	−50.50	7	114.99	125.68
	Hurdle	−50.63	7	115.26	125.94

Note. I display the numbers in this table to two decimal points, so that anyone can verify the values of AIC and BIC by calculation (e.g. *Poisson AIC* = −2 × (−233.64) + 2 × 4 = 475.28). However, differences smaller than 1 are unlikely to be meaningful for AIC and BIC, so that it's entirely appropriate to round them to whole numbers when reporting and evaluating them. For example, the difference in AIC between the zero-inflated negative binomial and hurdle negative binomial models is so small as to be negligible. The 'no excess zeroes' models have predictors *Population1000*, *Vendors1000* and *Lower10*. The zero-inflated and hurdle models have predictors *Population1000*, *Vendors1000* and *Lower10* in the logistic part, and predictor *Lower10* in the count part of the models. AIC = Akaike information criterion; BIC = Bayesian information criterion.

From Table 5.10, we can see four things:

1   A negative binomial distribution seems more appropriate than a Poisson distribution. We see this because both AIC and BIC are always smaller for the negative binomial models than for their Poisson counterparts (e.g. the negative binomial hurdle model has lower AIC and BIC values than the Poisson hurdle model).
2   Accounting for excess zeroes seems to improve the models. We see this because the zero-inflated and hurdle negative binomial models have smaller AIC and BIC values than the plain negative binomial model.
3   The zero-inflated and hurdle negative binomial models are approximately equivalent. We see this because they have more or less the same AIC and BIC values.
4   Overall, AIC and BIC suggest the same conclusions in this case.

These results suggest that either the zero-inflated negative binomial model or the negative binomial hurdle model is most appropriate. I present the results from the hurdle model with negative binomial count distribution as an example in Table 5.11. The results are very similar to the hurdle Poisson model for the same data (see Table 5.7). In fact, the estimates and standard errors for the hurdle part are the same. The hurdle part of the model only predicts whether the outcome is zero or not, so it is not affected by the decision whether the non-zeroes are modelled as zero-truncated negative binomial or zero-truncated Poisson. But in the zero-truncated negative binomial part, there is a difference: the *IRR* for *Lower10* in the negative binomial hurdle model is a bit smaller than in the Poisson hurdle model, and the standard error is larger, so that the confidence interval is wider. This suggests that the Poisson hurdle model overestimated the precision with which we were able to estimate the effect of *Lower10*. The negative binomial hurdle model in this case leads us to be cautious about claiming an effect of *Lower10* on the number of police operations, after accounting for zeroes: the confidence interval suggests that the data are consistent with a negative, positive or zero effect of *Lower10* on the number of police operations, once zeroes have been taken into account by the hurdle part. This shows the benefit of choosing a well-fitted model: if we had used the Poisson model, which is probably not appropriate, we would have arrived at an overoptimistic conclusion about the relationship between *Lower10* and the number of police operations (after accounting for structural zeroes).[xiv]

Table 5.11 Police operations in Santiago districts: hurdle model with negative binomial count distribution

Zero-Truncated Negative Binomial	Coefficient	SE	IRR	95% CI
Intercept (β_0)	0.779	4.718		
Lower10 (β_1)	−0.238	0.306	0.79	[0.43, 1.44]
α	14.085			
Hurdle Model	**Coefficient**	**SE**	**OR**	**95% CI**
Intercept (γ_0)	1.349	2.004		
Vendors (γ_1)	0.024	0.470	1.02	[0.41, 2.58]
Population (γ_2)	0.270	0.119	1.31	[1.04, 1.65]
Lower10 (γ_3)	−1.149	0.443	0.32	[0.13, 0.76]

Note. SE = standard error; IRR = incidence rate ratio; CI = confidence interval; OR = odds ratio.

Some general comments on AIC and BIC

AIC and BIC are measures of relative model quality. They are not comparable across different data sets, or across models with different outcomes. This is because the value of the AIC or BIC (e.g. 125.94 for the BIC of the negative binomial hurdle model applied to the Santiago data) does not have any interpretation by itself. It is only in comparison with other models that this number becomes meaningful.

A difference of less than 1 between two AIC or two BIC values is generally considered negligible. Unfortunately, there is no clear cut-off point above which a difference is considered meaningful. When comparing more than two models, the relative

---

[xiv]Note, however, that the results from the hurdle part in Table 5.11 suggest that *Lower10* is negatively associated with the probability of observing a non-zero, controlling for the number of vendors and population size. This suggests that poorer districts are more likely to have zero police operations than richer districts. So the results still support Holland's research hypothesis that in a politically decentralised city like Santiago, there will be fewer police operations in poorer districts.

sizes of the various pairwise differences can help to guide interpretation. For example, in Table 5.10, AIC and BIC indicate that the zero-inflated and hurdle negative binomial models are essentially very similar in quality but that both are much better than any of the other models considered.

You can use AIC and BIC for the comparison of nested or non-nested models. There is nothing wrong in principle with using them even in situations where a hypothesis test would also be appropriate. If you use both AIC and BIC, often both will suggest the same conclusions (as in Table 5.10), but sometimes they may not. The AIC has a tendency to favour the larger model more often than the BIC (because in BIC, the penalty term for additional parameters is larger than in AIC). There is no general rule to tell you what to do if AIC and BIC disagree. Some analysts therefore decide to use either one or the other. This decision should be made before seeing the data.

A final remark on comparing non-nested models: in the past, a hypothesis test called the Vuong test was often used for comparing zero-inflated models to ordinary Poisson or negative binomial models. However, the Vuong test has been shown to be inappropriate for this purpose, so I do not recommend using it in this context. Box 5.4 gives more information.

Box 5.4

The Vuong Test: A Warning

In the past, some textbooks recommended the use of a test called the Vuong test for comparing count models without zero-inflation with zero-inflated or hurdle models. The Vuong test is a hypothesis test designed for comparing non-nested models. It is a perfectly good test for some modelling situations, but unfortunately it is not appropriate for investigating whether zero-inflation or hurdle models improve prediction relative to ordinary count models (Poisson or negative binomial). This has been shown by Wilson (2015). So you should not use the Vuong test for deciding whether to use a zero-inflated (or hurdle) count model.

Inference for individual parameters and nested models within the same model type

Once we have decided which type of model we deem appropriate (e.g. Poisson or negative binomial, zero-inflated or not), we may be interested in confidence intervals and significance tests for individual slope coefficients of our predictors. We may also wish to compare nested models that differ by more than one parameter. Let's now consider these issues formally.

As is the case with all models discussed in this book, software packages calculate standard errors for all coefficients, have options for displaying confidence intervals and (usually by default) provide statistical tests for individual coefficients. For the models in this chapter, it is generally true to say that the default standard errors and confidence intervals provided by software can be trusted if the assumptions underlying the model all hold.

For example, a 95% CI around a coefficient is calculated as

$$CI_{0.95} = \hat{\beta} \pm 1.96 \times SE_{\hat{\beta}}$$

where $\hat{\beta}$ is a coefficient estimate, and $SE_{\hat{\beta}}$ is its estimated standard error. For confidence intervals with different coverage probabilities (other than 95%), the number 1.96 in the above equation would be replaced by the appropriate quantile from the normal distribution. For example, for a 99% CI, we would use 2.58. All results tables in this chapter display 95% CIs, but this is an arbitrary (albeit conventional) choice. Confidence intervals for IRRs are obtained by exponentiating the bounds of the confidence intervals for the coefficients.

The confidence interval as defined above relies on the assumption that coefficient estimates are normally distributed across different samples. This will be approximately true when all model assumptions hold, and when the sample size is sufficiently large. As an alternative, profile likelihood confidence intervals (as explained in Chapter 2 for logistic regression) can also be calculated.

A z-test of an individual coefficient is also based on the normality assumption. The z-statistic is

$$z = \frac{\hat{\beta} - \beta_0}{SE_{\hat{\beta}}}$$

where $\hat{\beta}$ and $SE_{\hat{\beta}}$ are defined as above. The z-statistic is evaluated against a normal distribution to obtain a p-value. The null hypothesis of this test is that $\beta = \beta_0$. Often the hypothesis of interest is that $\beta = 0$, so that the z-test statistic becomes $z = \hat{\beta} / SE_{\hat{\beta}}$. This test is the default displayed in most software packages.

Like the confidence interval, the z-test is valid if all model assumptions hold and the sample is sufficiently large. An alternative is to use likelihood ratio tests. For example, we might use either the z-test or the likelihood ratio test to evaluate the strength of the evidence against the null hypothesis that the number of street vendors is not related to the number of police operations in Lima districts, after controlling for district population size and poverty level. The likelihood ratio test also relies on model assumptions being met, but unlike the z-test, it is valid in small samples.

Likelihood ratio tests can also be used to compare nested models that differ by more than one predictor variable. The application of the likelihood ratio test for this purpose is no different to its use in logistic regression. Both z-tests and likelihood ratio tests have been introduced in Chapter 2, with further examples given in Chapters 3 and 4. So I won't give another example here.

On standard errors in count models

Most statistical software packages by default give standard errors for coefficient estimates of count models. These default standard errors are appropriate if all model assumptions hold. However, often we are not entirely certain whether all model assumptions are completely met (see also the section 'Model Evaluation'). This is why some authors recommend using different standard errors instead, the so-called *robust standard errors*. Robust standard errors give us an appropriate measure of the precision of our estimates even if model assumptions are violated. For example, if we fitted a Poisson regression to data that really follow a negative binomial distribution, the default standard errors would likely be incorrect. They would likely be too small, giving us an overly optimistic view of the precision of the estimates, such that confidence intervals would be narrower than they should be, and statistical tests would wrongly lead us to reject the null hypothesis too often. Robust standard errors, on the other hand, would be appropriate despite having fit the wrong model. So with robust standard errors, we would obtain accurate confidence intervals and z-tests about our coefficients, despite violation of model assumptions.

On the other hand, if model assumptions actually hold, robust standard errors may be larger than they need to be, thus giving us an overly pessimistic view of the precision of our estimates. Confidence intervals would be wider than they need to be, and we would lose statistical power in z-tests: we would wrongly fail to reject the null hypothesis more often than with the default standard errors. So ultimately, whichever standard errors we choose, there is a risk of making the wrong choice.

There are two types of robust standard errors: standard errors based on the sandwich estimator and bootstrap standard errors. These are not unique to models for count data, however, and so I will not discuss them in detail in this chapter. (See Chapter 6, section 'Beyond This Book', for more information and a reference.)

Model evaluation

As with any statistical model that we estimate on a data set, we should investigate the fit of count models visually, in order to investigate model assumptions and check for outliers. To this effect, we employ residual diagnostics similar to those for linear regression (see *The SAGE Quantitative Research Kit*, Volume 7). However, the type of residual we shall employ in the context of count models is different from the standardized residuals that we use for linear regression.

Deviance residuals for Poisson and negative binomial regression

We shall employ **deviance residuals** for Poisson and negative binomial regression models. Deviance residuals are calculated differently for different models. (There is a general formula, which is the same for all models that belong to the class of *generalized linear models*, but it's too complicated to explain in this book.) You can think of deviance residuals as a type of standardized residual. When the observed count matches the prediction exactly, the deviance residual is zero. When the observed count is larger than our model prediction, the deviance residual is positive. When the observed count is smaller than the prediction, we get a negative deviance residual. The larger the difference between the observed and predicted count, the larger the deviance residual (in absolute terms).

If the model is well fitted, the deviance residuals should have

- a symmetric distribution, and the mean should be close to zero;
- the same variance across the range of predicted values (homoscedasticity),[xv] and
- no pattern suggesting non-linearity, when plotted against the logarithm of the predicted values.

Furthermore,

- in large samples, the deviance residuals should approach a normal distribution, and
- we can use the deviance residuals to check for outliers and influential observations.

These criteria are very similar to those used in, Volume 7 of *The SAGE Quantitative Research Kit* when evaluating linear regression models by investigating standardized residuals.

Let's have a look at an example. Consider once again Model 5.2, the negative binomial regression of police operations in Lima. To carry out residual diagnostics, we

---

[xv]Note that Poisson and negative binomial regression do not assume homoscedasticity of the prediction errors. In both Poisson and negative binomial regression, the variance of the prediction errors depends on the predicted mean, so the variance is expected to be different for different predicted values. However, in the calculation of deviance residuals, this heteroscedasticity is adjusted for, such that the deviance residuals should indeed be homoscedastic if model assumptions hold.

calculate the deviance residuals (using statistical software) and describe their distribution. The deviance residuals for Model 5.2 are described in Table 5.12 and illustrated in Figure 5.13.

Table 5.12 Distribution of deviance residuals from a negative binomial regression of police operations in Lima districts (Model 5.2)

Minimum	−2.71
Median	−0.26
Mean	−0.29
Maximum	1.44
Standard deviation	1.09

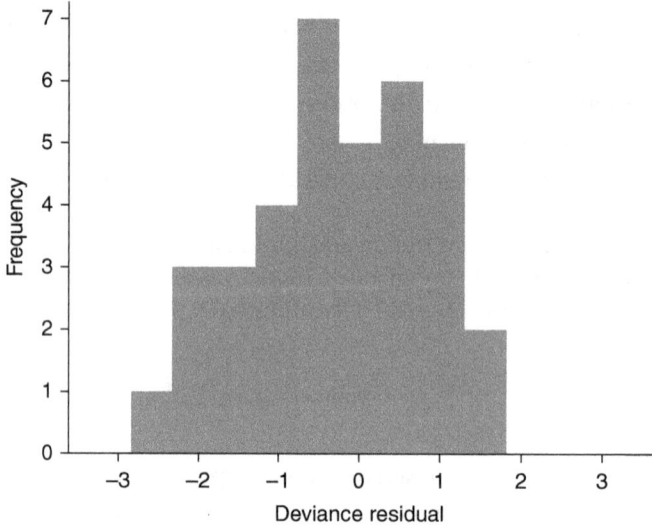

Figure 5.13 Histogram of deviance residuals from a negative binomial regression of police operations in Lima districts (Model 5.2)

Table 5.12 shows that the deviance residuals have a slight negative skew. The mean and median are negative, and the largest negative residuals are larger than the largest positive residuals. The skew can also be seen visually in the histogram (Figure 5.13).

An important diagnostic plot is the spread-level plot, a scatter plot of deviance residuals against the logarithm of the predicted outcome. We are plotting the logarithm of the predicted outcome, because this is the transformation that we have used for the outcome in our model. A spread-level plot for Model 5.2 is shown in Figure 5.14.

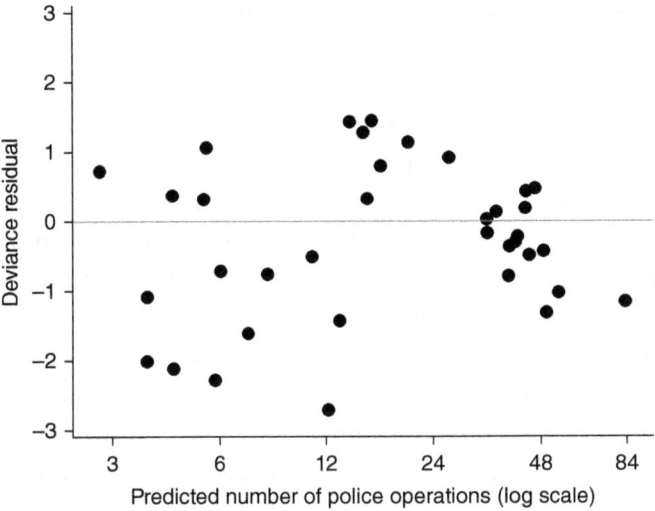

Figure 5.14 Spread-level plot of deviance residuals against log-predicted police operations from a negative binomial regression on the Lima data (Model 5.2)

The criteria for evaluating this plot are the same as shown in Volume 7 of *The SAGE Quantitative Research Kit* for linear models (except that we are not concerned with checking for normality, at least not in small samples). In our example, the situation is a bit unclear. The deviance residuals appear to follow roughly an inverted U-shape. This would suggest that we could try taking the square transformation of one of our predictors to see whether this removes the pattern. We could plot each predictor variable against the deviance residuals to find out which predictor should be transformed. On the other hand, with small samples like this one (36 districts), we should be cautious not to overinterpret small deviations from expected random scatter of the deviance residuals. Refer to Chapter 3 in Volume 7 of the *Kit*, for some ideal typical situations for 'nice' and 'definitely not nice' residual patterns.

In general, what should we do if we find that there is heteroscedasticity or non-linearity in the deviance residuals? The potential remedies to try are the same as for linear regression, or for a poorly calibrated logistic regression model:

- Check for outliers and influential observations
- Check whether you have failed to include an important predictor variable
- Check whether you have failed to include an important interaction
- Check whether one or several of the predictors may need to be transformed
- Check whether a different transformation of the outcome (i.e. a different link function, see Box 5.2) might be appropriate

Offsets: accounting for population size, time of exposure, or area

To complete this introduction to modelling count data, there is one more issue we need to consider. Recall that all the count models we have discussed in this chapter rely on the assumption that the observed counts are comparable across the units of observation. For example, when we considered the number of deaths by horse kick per army corps, we specified that each count referred to the same observation period (1 year). If observations had been taken over different observation periods (1 year for corps A, 18 months for corps B, say, and so on), they would not have been comparable directly.

The same goes for the data on police operations in Latin American cities: the number of police operations should be measured over the same time period in all districts in a city to ensure an accurate comparison of districts. On the other hand, districts also vary in population size, and we took account of this by adjusting our models for population size. That is, we added population size as a predictor in the model and estimating its slope coefficient. This section will look at another type of adjustment, which is called an offset.

What is an offset, and why might we need one?

Offsets may be applied in many research situations where counts may not be directly comparable across units of observation, because they were measured over different time periods, sizes of areas, or population sizes. Consider, for example:

- The number of visits to the dentist per person, which may be reported over different time periods for different people. We may wish to adjust for observation time to account for this in our analysis.
- The number of cyclist deaths per city per year. Directly comparing cities with many cyclists to cities with few cyclists would bias the comparison. We would like to adjust the analysis by a suitable indicator of cycling frequency, such as total number of miles cycled in each city in a given year.
- The number of goals scored by a football player[xvi] over a league season. Players may have played in different numbers of matches, and may not have played over the whole 90 minutes in each match. So we may wish to adjust our model to take account of the number of minutes each player spent on the pitch over the season.

An offset is a predictor without a coefficient, which is included in the model purely as an adjustment. The general form of a count model with an offset is

$$\log(\mu_i) = \beta_0 + \beta_1 x_{1i} + \beta_2 x_{2i} + \ldots + \log(exposure_i)$$

[xvi]Or 'soccer player', if you're used to the American term.

where log(*exposure*$_i$) is called the offset, and *exposure* is the time period, population size or area size by which we wish to standardize the expected count μ_i. Note that the term log(*exposure*$_i$) does not have a coefficient.

You can use an offset with any count model discussed in this chapter, including Poisson regression and negative binomial regression, as well as zero-inflated, hurdle and zero-truncated models. Model comparisons and model evaluations can and should be conducted for models with an offset in the same way as for models without an offset.

Research example: socio-economic differences in the uptake rate of free eye tests

Let's look at an example of how to use an offset in practice. The example will also help to explain why an offset is defined in the way it is. Shickle and Farragher (2015) were interested in social inequalities in the uptake of free eye examinations. In England, preventive eye examinations are free for people under the age of 16 years, as well as for people over the age of 59 years. But for a variety of reasons, some people choose not to take up the offer of a free eye test. Shickle and Farragher investigated whether socio-economic status was related to the number of free eye examinations that were taken up. Socio-economic status was measured by an area-level variable, which classified each area by the Index of Multiple Deprivation (IMD). The areas were divided into five quintiles on the IMD, from IMD:I (most deprived 20% of areas) to IMD:V (least deprived 20% of areas). Box 5.5 explains the data set in more detail. Unusually for this book, I did not use real data from the original study, but simulated a data set that gave similar results as those published by the authors, and that will serve us well as a simple example.[xvii]

Box 5.5

Shickle and Farragher's Eye Examination Data

Shickle and Farragher (2015) analysed data on the number of free eye examinations conducted in the English city of Leeds in February and March 2011. Their data set consisted of the following variables:

(Continued)

[xvii]This is because the original data contained potentially sensitive information, so that the authors were unable to share them with me.

- *Area:* LSOA, or 'lower super-output area', a division of the city of Leeds with between 1000 and 3000 inhabitants each
- *Population size:* the number of people estimated to live in each area in 2011
- *Index of Multiple Deprivation:* this classifies each area according to indicators of deprivation, such as the number of people who receive state benefits, the number of people who are unemployed, measures of community health and others (for details, see UK Department of Communities and Local Government, 2011). IMD is recorded in this data set in quintiles, such that the first quintile indicates the most deprived 20% of areas, and the fifth quintile indicates the least deprived 20% of areas
- *Number of free eye tests:* number of tests in February and March 2011 for children under 16 years of age.

The data set thus looks something like Table 5.13.

Table 5.13 Hypothetical data set recording the number of free eye tests for children under 16 years per area, along with area population and IMD quintile group

Area ID	Population	IMD Quintile	Number of Eye Tests
1	1780	II	11
2	1267	II	10
3	1463	IV	6
4	1693	I	6
5	1275	V	4
6	1279	III	7
7	1570	I	8
⋮	⋮	⋮	⋮

Note. IMD= Index of Multiple Deprivation.

The particular data in this table are made up. I was unable to access the original data used by Shickle and Farragher (2015), so instead I simulated a fictitious data set that gives similar results to those published by the authors.

Let's first examine a summary of the data. Table 5.14 shows the total number of free eye tests, the total population, and the test uptake rate per 1000 people for each of the IMD quintiles.

Comparing areas in different IMD quintiles regarding the mean number of eye examinations, we see that the most deprived areas had the largest number of eye examinations. However, the most deprived areas also have the largest total population. Thus, if the aim is to see whether people in deprived or less deprived areas are more likely to do an eye test, we should consider the *uptake rate*, that is, the number

Table 5.14 Summary of hypothetical data illustrating number of eye tests in 100 areas

IMD	Number of Areas	Total Number of Eye Tests	Total Population	Uptake Rate: Number of Eye Tests per 1000 People
Most deprived: I	100	864	196,917	4.39
II	100	727	153,327	4.74
III	100	606	135,172	4.48
IV	100	653	135,641	4.81
Least deprived: V	100	715	132,619	5.39

Note. The uptake rate was calculated as follows:

$$\text{Uptake rate} = \frac{1000 \times \text{Mean number of eye examinations}}{\text{Mean population}}. \text{ IMD} = \text{Index of Multiple Deprivation.}$$

of free eye tests per 1000 people in the population, as shown in Table 5.14. Here we see that the uptake rate is actually lowest in the 'most deprived' areas, where 4.39 free eye tests were conducted per 1000 people. In comparison, 5.39 free eye tests per 1000 people were conducted in the least deprived areas.

How can we model the uptake rate? Following the logic of the models we have encountered so far, we might propose the model:

$$\log\left(\frac{tests_i}{pop.size_i}\right) = \alpha + \beta_1 IMD : II_i + \beta_2 IMD : III_i + \beta_3 IMD : IV_i + \beta_4 IMD : V_i$$

where

- areas are numbered $i = 1, 2, \ldots, n$,
- $tests_i$ is the number of eye tests conducted in area i,
- $pop.size_i$ is the population size of area i,
- $\frac{tests_i}{pop.size_i}$ is the uptake rate of area i, and
- *IMD:II, IMD:III, IMD:IV* and *IMD:V* are dummy variables identifying areas classified in the four IMD quintiles II, III, IV and V, respectively. The most deprived quintile (IMD:I) is the reference category.

This seems sensible, but there is a problem: the outcome, if defined in this way, is not a count variable. The uptake rate can take non-integer values, so by definition it won't follow either a Poisson or a discrete negative binomial distribution. So it seems as though we won't be able to use one of the models we have discussed so far. However, it turns out that we can employ a simple mathematical trick that will allow us to use a count model for these data after all.

The trick is based on knowing that, in general, $\log\left(\frac{a}{b}\right) = \log(a) - \log(b)$. Thus, $\log\left(\frac{tests_i}{pop.size_i}\right) = \log(tests_i) - \log(pop.size_i)$, and we can rewrite our model as follows:

$$\log(tests_i) - \log(pop.size_i) = \alpha + \beta_1 IMD : II_i + \beta_2 IMD : III_i + \beta_3 IMD : IV_i + \beta_4 IMD : V_i$$

The final step is to take the term log(*pop.size_i*) over to the right-hand side of the equation. This gives:

Model 5.5a:

$$\log(tests_i) = \alpha + \beta_1 IMD : II_i + \beta_2 IMD : III_i + \beta_3 IMD : IV_i + \beta_4 IMD : V_i + \log(pop.size_i)$$

Written in this way, Model 5.5a has a count outcome: the number of eye tests conducted in area *i*. The predictors are four dummy variables identifying the deprivation quintile of area i, and the predicted count is adjusted for the population size of area *i*. The term log(*pop.size_i*) is the offset. Note, once again, that the offset does not have a coefficient. We are not estimating the effect of the offset on the count outcome. Rather, we are assuming that the offset acts as an appropriate adjustment. Another way to express the same thing is to say that we set the coefficient of the offset to be equal to 1.

Model 5.5a would give us, as the predicted value, the predicted number of eye tests per person. This is perfectly adequate in principle, but since there are on average only about five eye tests for every 1000 people, the predicted values from Model 5.5a would all be very small. A convenient modification in such a case is to model the uptake rate per 1000 people instead. We can do this by adjusting the offset to be equal to $\log\left(\frac{pop.size_i}{1000}\right)$. Thus, we will estimate the following model:

Model 5.5b:

$$\log(tests_i) = \alpha + \beta_1 IMD : II_i + \beta_2 IMD : III_i + \beta_3 IMD : IV_i + \beta_4 IMD : V_i + \log\left(\frac{pop.size_i}{1000}\right)$$

Estimating Model 5.5b on the eye test data yields the results displayed in Table 5.15. The offset is shown in its own row, with coefficient equal to 1. Note again that this coefficient is *set* to 1; it is not estimated. The offset does not necessarily need to be reported in a results table, but I have included it here for clarity.

The estimated coefficients of the IMD quintiles and their confidence intervals indicate that there is probably little difference in the uptake rates between IMD quintiles I, II, III and IV. However, the least deprived areas (IMD Quintile V) are estimated to have a 23% higher uptake rate than the most deprived areas (95% CI between 11% and 36%). So it looks as though free eye tests are disproportionately taken up in areas where families are least likely to need them (because they could afford to pay) compared to areas where families are most likely to need them. Obviously, it would be possible in principle to add further predictors to the model, representing other characteristics of the areas, if information on those were available.

Table 5.15 Estimates from a Poisson regression of the number of eye examinations in 100 hypothetical areas

	Coefficient	SE	IRR	95% Confidence Interval
Intercept	1.478	0.034		
IMD Quintiles (ref: IMD:I, most deprived)				
II	0.078	0.050	1.08	[0.98, 1.19]
III	0.022	0.053	1.02	[0.92, 1.13]
IV	0.093	0.052	1.10	[0.99, 1.21]
V (least deprived)	0.206	0.051	1.23	[1.11, 1.36]
Offset: log(Population/1000)	1			

Note. IMD = Index of Multiple Deprivation; *SE* = standard error; *IRR* = incidence rate ratio.

Chapter Summary

Previous chapters in this book have, by and large, each introduced a single type of statistical model. This chapter is rather different: it has discussed eight different models for count data and has given advice on how to decide which one might be most appropriate for a given data problem and analytic aim. Sometimes we can base the decision about the most appropriate model type on our knowledge of the way the data have been collected or generated. For example, if we know that we could not have observed any zeroes, a zero-truncated model should probably be considered. Or, to give a different example, if we suspect that the zeroes have different predictors than the number of events among the non-zeroes, a hurdle model should probably be considered.

We have then considered how we can use a data set to compare different plausible candidate models. The boundary likelihood ratio test was recommended for deciding between a Poisson and a negative binomial model. We have also discussed information criteria as a way of comparing models that are not nested and have noted that information criteria can also be used for nested model comparisons, either instead of hypothesis tests or in conjunction with them.

There are other models for analysing count data, besides those discussed here. I mention two of them briefly here:

1   A *quasi-Poisson model* is a Poisson model with scaled standard errors to account for overdispersion. A quasi-Poisson model is therefore often used as an alternative to a negative binomial model, when overdispersion is present such that an ordinary Poisson model is not appropriate (Hilbe, 2014, p. 79f).

2   Another model for overdispersed count data is the *Poisson inverse Gaussian (PIG) model*. This is similar to a negative binomial model, but assumes a different outcome distribution. Hilbe (2014) states that the PIG model is likely to be useful if the overdispersion is very high, especially when the overdispersed count variable has a very high peak and a long tail to the right. Hilbe (2014, p. 164) gives the example of length of stay (number of days) in hospital, where most patients are discharged very quickly (the high peak), but some stay on for a very long time (long tail, and hence large variance). Hilbe suggests that a PIG model might fit such data better than a negative binomial model.

Further Reading

Hilbe, J. M. (2014). *Modeling count data*. Cambridge University Press.
This book gives a thorough overview of statistical models for count data, discusses (in greater depth) all the models introduced in this chapter, and also covers several models and techniques for which there was no space here. Hilbe's book demands a higher degree of mathematical understanding than the book you are reading now, but it includes helpful advice on how to fit count models in R and Stata, with illuminating examples.

Long, J. S., & Freese, J. (2014). *Regression models for categorical dependent variables using Stata* (3rd ed.). Stata Press.
This book is a less mathematical treatment and also includes instructions on how to use Stata to analyse count data.

6

THE PRACTICE OF MODELLING

Chapter Overview

Statistical models aim to separate the signal from the noise (Silver, 2012). When we fit a model to a data set, our objective is to gain insight into the relationships between social phenomena from the traces they leave in the data we see. When we evaluate our model, we also pay attention to the random variation and measurement errors that add noise to our data sets, and we try to diagnose whether our assumptions about the noise (e.g. about the residuals) are correct or not, whether we may have misinterpreted some noise as a signal, or whether we may have missed a signal amid the noise.

A good statistical model should have three characteristics:

1   *Fit to the data:* the model should not be demonstrably inconsistent[i] with the data.
2   *Scientific plausibility:* the model should be plausible based on what we know from other evidence about the subject we are studying.
3   *Parsimony:* the model should be as complicated as necessary to achieve the two aims above, but as simple as possible.

That sounds very nice. It's a proud moment when, at the end of a piece of data analysis, we can announce that our model fits the data well, is parsimonious and yields results that shed light on a new scientific insight while at the same time being compatible with well-evidenced knowledge in our field. But how do we arrive at such a Good Statistical Model?

One purpose of this book is to serve as a helpful companion in studying essential techniques of regression models for categorical and count outcomes. That is what Chapters 2 to 5 are chiefly devoted to. This final chapter concentrates on another purpose: to consider the general principles of good statistical modelling in practice. What I say in this chapter is meant to apply to all the models discussed in this book, as well as to regression models more generally.

I invite you to consider what you want from a statistics textbook. Do you expect the book to tell you what you, as a data analyst, should do? Sometimes it might indeed tell you what to do; for example, I have recommended certain statistical hypothesis tests over others. However, this book also has a more fundamental aim: to help you learn enough about the principles of statistical modelling to enable you to think through a research problem for yourself and make your own reasoned decisions about how to conduct your data analysis.

Decision-making in statistical modelling

Statistical data analysis is sometimes taught as a set of rules:

- 'If the *p*-value is below 0.05, reject the null hypothesis';
- 'calculate a confidence interval for all your estimates';

---

[i] I learned the phrase 'not demonstrably inconsistent' from Peter Diggle (2019).

- 'use linear regression for normally distributed outcomes';
- 'if the residuals display a non-linear pattern, transform your variables';
- 'if the outcome is dichotomous, use logistic regression'
- and so forth.

Students might be tempted to come to the conclusion that being good at statistics involves knowing the right rules, and that by following these rules they can be sure to be doing the correct statistical analysis. The typical statistics assessment in the form of an exam, where you get points for the 'correct' answers, reinforces this view. So do some adverts for statistical software, as Gelman and Hennig (2017) point out:

> Often . . . users of statistics . . . expect that for their data analysis task there is a unique correct statistical method. This expectation is exploited by the marketing strategies for some data analysis software packages that suggest that a default analysis is only one click away. . . . Decisions that need to be made are taken out of the hand of the user and are made by the algorithm. (p. 969)

Yet good statistical data analysis is not achieved in this way. Rules do not provide all the answers, and software does not solve all your problems. Four considerations are important:

1 For a statistical model to be useful, there needs to be a good match between the model and the aims of the statistical investigation. No algorithm can ensure such a match.
2 Many 'rules' stated in textbooks should not be understood as prescriptions to be followed without thinking, but rather as guidelines, as advice that may apply in many situations, but not necessarily always.
3 There is not necessarily a single correct or single best answer to the question 'Which model should I choose?'
4 Statistical modelling involves making decisions that partly depend on the judgement of the analyst.

In the following, I elaborate on each of these points.

Match between the statistical model and the aims of the research

Statistical modelling starts with an aim: a research question we would like to answer. The first consideration when specifying a statistical model is therefore the model's capability of addressing the goals of the investigation. This requires thought. Each investigation, each data set, each research aim is different and needs to be considered thoroughly. No rule book and no software can do this job for the researcher. Two researchers who analyse the same data set, but who pursue different research questions, might use quite different statistical models.

For example, consider short-term mortality after emergency surgery. This is often measured as 30-day mortality: whether a patient dies within 30 days from the date of surgery or whether she survives. For emergency bowel surgery in the UK, the 30-day mortality rate is about 10% (Eugene et al., 2018). This is a high mortality rate, because when patients come to hospital for an emergency bowel surgery, they are usually very ill. But the risk of death varies considerably between patients, depending on how ill they are. The patient's physiological status is assessed via medical tests prior to the operation, which measure variables such as blood pressure, the amount of white blood cells and so forth. Let's call these variables physiological risk factors.

Now, suppose you wish to study whether some hospitals achieve better survival rates than others. Then, it would be important for you to statistically adjust for the possibility that hospitals differ in their patient case mix. In other words, one hospital might treat patients who are on average sicker than the patients of another hospital. For a fair comparison of the mortality rates of those hospitals, it is desirable to adjust for physiological risk factors. The statistical adjustment would allow you to compare the survival chances of two patients *with the same risk status*, if they went to two different hospitals. The development of a risk adjustment model for emergency bowel surgery, in order to facilitate the comparison of mortality rates between hospitals, is described in Eugene et al. (2018). The model employed in this study is a logistic regression of 30-day mortality on a large number of risk factors.

Now consider a different question: is the distance a patient has to travel to get to hospital for an emergency surgery related to that patient's chances of survival? This is an important question, because the answer might inform organization and planning of healthcare services: is it better to have many small local hospitals, such that travelling times are short, or is it preferable to have fewer but larger hospitals, because travelling times don't matter, but larger hospitals have other advantages? Now, one of the potential consequences of having far to travel to hospital for an emergency surgery could be that the patient gets sicker on the way. So, hypothetically at least, the patient's physiological status at the time of the operation may partly be a consequence of the patient's travel time. Then, in order to estimate the effect of travel times on the risk of death, it is important *not* to adjust for physiological risk factors, because doing so would 'adjust out' one of the ways through which a long travel time might contribute to an increased risk of death. Salih et al. (2020) conducted a study of the relationship between distance to hospital and risk of death after emergency bowel surgery, using the same data set as Eugene et al. (2018). The model was a logistic regression of patient 30-day mortality on distance to hospital, but *not* adjusting for physiological risk factors (although Salih et al., 2020, did adjust for some patient characteristics, which were thought not to be influenced by travel time).

So here is an example of two studies using the same data set, studying the same outcome with the same model type (logistic regression) but using different predictors and making different decisions regarding which variables to statistically adjust for. These differences are entirely appropriate, because the two studies pursue different aims. In both cases, the decision regarding how to specify the statistical model was made after carefully considering the research question and the process that each study wished to investigate. Following a decision rule such as 'include all variables whose coefficients are statistically significant' would have been entirely useless as a way of arriving at a meaningful model.

Most rules are just guidelines

Next, let's consider how to think about advice given in statistical textbooks, especially books written for students whose first degree is not statistics. Such books tend to focus on the application of statistics (such as this one does), rather than on the underlying theory and mathematics. Rules given in such books are useful in providing a general orientation, but that does not mean that you should follow them invariably or without thought. As you learn more and delve deeper into statistics, you have a chance to develop an understanding that will allow you to choose from a wider range of options, and be a more flexible analyst who is able to cope with a wide variety of situations.

Consider, as an example, the choice of an appropriate model for a dichotomous outcome. Chapter 2 of this book introduces binary logistic regression for this purpose, and only cursorily mentions alternatives. The chapter also states that the use of linear regression for binary outcomes – the *linear probability model* – has disadvantages that make it likely to be inappropriate in many cases. While this statement is not wrong, it is also true that the linear probability model can have some advantages over binary logistic regression. In particular, the linear probability model yields coefficient estimates that can be interpreted as effects on the average probability of the outcome, rather than as effects on the odds. This can make coefficients more interpretable for some applications. Thus, when none of the disadvantages of the linear probability model apply, it can be preferable to logistic regression (Battey et al., 2019). In other words, although the rule 'when your outcome is dichotomous, choose binary logistic regression' is a sound guideline that is unlikely to lead you far astray, there may be situations when it is better to discard this rule and make a different choice.

Usually you can't be sure that you have found the 'best' model

For any given data set and research question, there is an infinite number of models that can be demonstrated to be almost certainly wrong. Thus, for a person armed with powerful statistical software, but no knowledge of statistical modelling, there are infinite ways to produce erroneous findings. This is why it's important to pay attention to the rules: if you simply ignore them, you might end up with a model that is horribly wrong and gives you grossly misleading results. For example, consider Figure 6.1, which illustrates the average life expectancy and gross domestic product (GDP) per capita in 88 countries, measured in 2007. To model this relationship as a straight line (what Figure 6.1 calls 'naive linear regression') clearly leads to misleading results. Inspection of residual diagnostic plots from a linear regression using these data would confirm this (these data and this analysis were discussed in detail in *The SAGE Quantitative Research Kit*, Volume 7, Chapter 3). But if I ignore basic rules of good analysis and neither look at a plot of the data, nor conduct residual diagnostics, I may well accept this model. I would be wrong to do so.

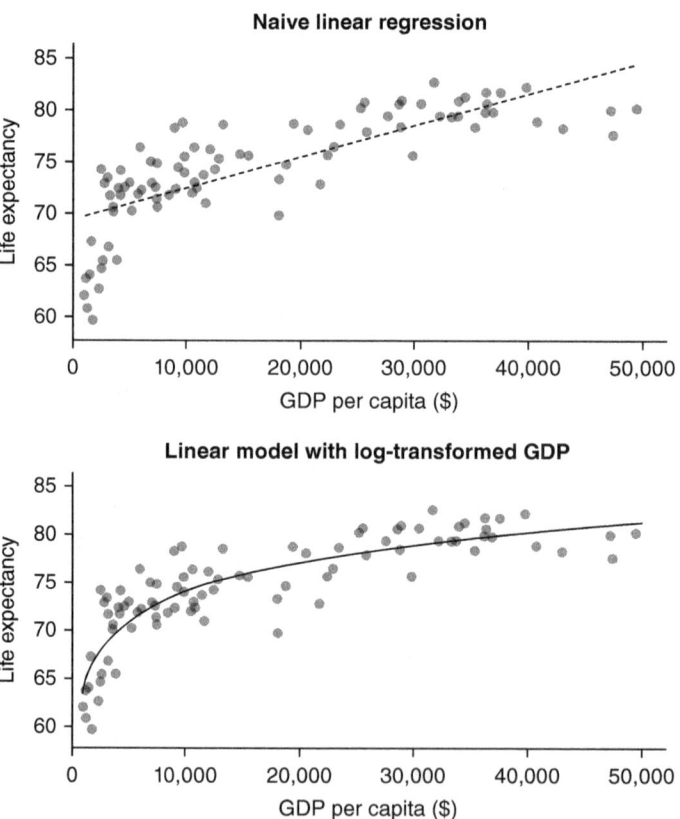

Figure 6.1 Linear regression of life expectancy on GDP per capita in 88 countries (2007): naive regression versus regression with log-transformed GDP

Note. The same data are shown in both plots. GDP = gross domestic product. Data were taken from Gapminder Foundation (Bryan, 2017). See www.gapminder.org

However, unfortunately it is not always true that, after eliminating all the wrong models, just a single model is left. Nor is it always clear whether the model that seems least wrong is actually correct. To stick with the example of life expectancy and GDP, although log-transforming GDP clearly leads to a better model than the naive linear regression (see Figure 6.1, again), we cannot be sure that log transformation is 'best'. There may well be other models that fit the data better, and that may also be better at predicting the life expectancy in countries not included in our analysis data set, or better at predicting a country's average life expectancy in the future. More generally, for any research problem, there may well be several plausible models, and the data at hand don't always allow you to decide which one is 'best', at least not with a high degree of certainty. In other words, in statistical modelling, there are a thousand ways of getting it wrong, but there is no way to be certain that you got it right!

Even if your data suggest one model to be clearly superior to all plausible others, this model is probably not an exact representation of the social process you are trying to understand. In the famous words of the influential statistician George Box,

> The most that can be expected from any model is that it can supply a useful approxima-tion to reality: All models are wrong; some models are useful. (Box et al., p. 440)

The London underground map is a model of London. If you try to use it to determine how far it is to walk from Baker Street to Paddington, you will find that the model is wrong. However, the map is useful if you want to travel on the underground and need to know which train to take. Statistical models are the same. To use them and to interpret them, you need to know what purposes they are good for.

The importance of the analysts' judgement

Statistical modelling involves making decisions. These decisions should be based on sound statistical principles, such as those explained in this book: confidence inter-vals, likelihood ratio tests, information criteria, investigation of residuals and out-liers, and so forth. Equally important are considerations about the context of the research question and the data: our modelling decisions should be justified based on our understanding of the topic we are investigating and our knowledge of how the data were collected. And they should be made, of course, without regard to whether the findings support a hypothesis that we like. Nonetheless, despite the emphasis on principles, sound understanding and objectivity, decisions that we make during sta-tistical data analysis often have an element of subjectivity, in the sense that a differ-ent analyst may have justifiably made a different decision (Gelman & Hennig, 2017).

To give a few examples, consider decisions such as the following:

- *Example 1 – Outliers:* Imagine that you have a data set with an outlier, which is an unusual observation and is moderately influential on the slope of a regression line. But the outlier is an important member of the sample, and its measurements are all valid. Should you exclude the outlier from your data set or not?

- *Example 2 – Transformations:* Imagine that you investigate the shape of the relationship between a predictor and an outcome. Using a square transformation of the predictor slightly improves the fit of your model compared to not using a transformation, but there is no scientific reason to think that the relationship should be anything other than linear, and so the slight improvement in fit with the square term might just be a chance occurrence. Should you use the square transformation or not?
- *Example 3 – Combining categories:* Imagine you are analysing an ordinal outcome variable with three categories. You are considering using an ordinal logistic regression. However, the middle of the three categories has a much smaller number of participants than the other two categories, and you are concerned that small cell sizes might make the coefficient estimates from ordinal regression unreliable. Should you combine the middle category with one of the other two, so that you end up with a binary variable and can use binary logistic regression? But if you do so, which of the other two categories should you combine the middle category with?

Such decisions can be difficult to make. Sometimes data exploration or the research context will point very clearly in one direction, but at other times they may not. Ultimately, the decision you take may depend on a subjective weighting of the advantages and disadvantages of each option. To stick with the first of the examples given above, one analyst might prefer to exclude a potential outlier in order to make sure that the model adequately represents most of the observations in the data; another analyst might prefer to include the outlier, so that the results can be generalised to the whole population from which the data were sampled. The decision may depend on the purposes of the investigation, the research question and the analyst's understanding of the underlying theory and empirical evidence from other studies in the same topic area.

In the end, whatever model we choose will always be open to critique, for example, because our analysis is

- based on measurements that may contain biases and inaccuracies;
- based on samples that may not be truly random, or not fully representative of the population we are investigating;
- based on data sets that may fail to contain some important predictors of our outcome;
- based on assumptions that, although hopefully plausible, may not be proven with certainty to be true, and at any rate may only be true approximately;, and
- based on analytic decisions that, while justifiable, may not be the only choices that could reasonably have been made.

Being open to critique is not a bad thing. It is a good thing. It means that we are open about how we arrived at our conclusions, and allow others to assess how valid these conclusions are. If our research report hides the modelling process from our readers, they can neither learn from our methods nor spot mistakes or problems, if there are any. Hiding the subjectivity of some of our modelling decisions might make our research report look more 'objective', but ultimately it harms scientific progress.

Some general principles that apply most of the time

Statistical modelling is an art as well as a science (Krzanowski, 1998, p. 243). The ability to make good data analytic decisions can improve with experience. Some general principles that often help and will rarely lead you astray are as follows:

- Consider the research question, or aim of your investigation, before you start your analysis.
- If possible, specify a plausible model before you see the data.
- Explore your data, and plot the data before you model them.
- Carefully think about how each variable was measured, how good an indicator each variable is for the concept you hope it is measuring, and whether there is any risk of bias or other errors due to the way the data were collected.
- Carefully think about the population or process you wish to study, and whether the data can be considered a random sample from that population or process; carefully consider the possibility that your sample or data set might be biased in some way, for example, due to systematic selection of certain cases over others.
- Talk to colleagues and other researchers, especially if they have been involved in collecting the data, have modelled similar data before, or are experts in the scientific field in which you are working. Consult about your modelling decisions and elicit others' views on the challenges you are facing.
- Be aware of model assumptions and use residual analysis to assess whether your assumptions are contradicted by the data.
- Plot the predicted values from the model against the data to allow you (and the readers of your research report) to see what the model implies and how much variation there is in the data around the predictions.
- Use confidence intervals as indicators of the uncertainty in your estimates and to gauge the range of plausible values for the parameters you are estimating.
- Use hypothesis tests only to evaluate important research questions or to inform important modelling decisions; don't use them as a default for every coefficient you estimate and don't use them in situations where they are not valid (e.g. if you've developed the hypothesis after you've seen the data).[ii]

Whatever you do, it is good practice to carefully document all modelling decisions that you made on the way to arriving at your final model, and to include an account and justifications of these decisions in your research report. This allows your readers to judge how far your results might be influenced by the decisions you have made. A useful technique to probe such influences yourself is sensitivity analysis, which we turn to next.

[ii]For more thoughts on statistical hypothesis tests, see the section 'Is Science in a Statistical Crisis? On *p*-Values and Hypothesis Tests' later in this chapter.

What if we're not sure about model assumptions: sensitivity analysis

All statistical models are based on assumptions, and inferences based on a statistical model are valid only if its assumptions are met. In practice, we are often not sure whether an assumption has been completely met, or whether an alternative assumption may be more appropriate. For example, we may be doubtful whether the relationship between a predictor and an outcome is linear or not, whether an outcome follows a Poisson distribution or not, whether an outlier ought to be excluded or not, or whether some outcome categories ought to be combined or not. In such situations, it is often useful to probe how sensitive our results are to the model assumptions we are making. The process of doing so is called sensitivity analysis.

A sensitivity analysis proceeds as follows:

- Specify alternative assumptions that you could plausibly make about the data.
  - Example: the count outcome is equidispersed versus the outcome is overdispersed.
- Specify the models appropriate under each assumption.
  - Example: Poisson regression versus negative binomial regression.
- Fit all models to the data.
- Compare results to judge how sensitive your conclusions are to making different assumptions.

For example, Holland (2015) analysed police operations against street vendors in Latin American capital cities (see Chapter 5). Her outcome, the number of police operations, was a count variable, but it was not clear whether it was equidispersed or not. If it was equidispersed, Poisson regression would be appropriate. On the other hand, if the outcome was overdispersed, negative binomial regression might provide a superior model. In the event, Holland analysed the data using Poisson regression,[iii] but also conducted a sensitivity analysis: she repeated her analysis using negative binomial regression instead of Poisson regression, keeping all other aspects of her models the same (Holland, 2015, p. 362). This allowed her to probe whether her conclusions were the same or different, when she made different assumptions about her outcome. In Holland's (2015) case, Poisson and negative binomial regression led to very similar results. This allowed her to argue that her conclusions were valid under either assumption.

In general, there are two types of conclusions from a sensitivity analysis: either the results do not differ (very much) with different assumptions, in which case we say

---

[iii]In fact, Holland (2015) used Poisson regression with robust standard errors (see the section 'Beyond This Book: Other Types of Models'). The robust standard errors allowed her to argue that her model inferences were valid even if the equidispersion assumption was violated.

that the results are *robust* to making different assumptions. Or the results do differ (considerably), in which case we say that the results are *sensitive* to the assumptions made. A result of the second type may stimulate further research.

How to test a finding: replication and out-of-sample prediction

Humans are good at generating meaning. We can always come up with an explanation of an event after it has occurred. Such an explanation-after-the-fact is called a *post hoc explanation*. Whether a post hoc explanation is correct or not is often difficult to decide.

In contrast, if you believe that you understand how something works, you can put this understanding to the test by making predictions about what is going to happen. If your predictions prove correct, this makes it plausible that your understanding is sound. An example of this is the scientific research on human-induced climate change. The statistical models of the Intergovernmental Panel on Climate Change are convincing for many reasons, including the depth of the scientific knowledge from many disciplines that they incorporate. But among these reasons is also the fact that the statistical predictions of future climate change implied by these models have been confirmed by the reality. The rise in global average temperatures from 1990 to 2012 has been consistent with the predictions of the Intergovernmental Panel on Climate Change climate model published in 1990 (Cubasch et al., 2013, p. 131). On the other hand, those who in 1990 did not believe that global temperatures would rise further than they already had by that point, have been proven wrong.

At the point in time at which I devise a statistical model, I usually do not have data on which I can test my predictions. I only have the initial data set, on which I estimate my model. When developing a model, one temptation is to try to make it fit to the currently available data set as tightly as possible. But there is a catch: if I make my model complicated enough, I can always make it fit the data perfectly. This is the statistical modelling equivalent of a post hoc explanation. However, a perfectly fitted model usually does not yield useful insights, and it is likely to be useless when tested on new data. A good model, in contrast, should be able to predict new data that were not used in its development. This is called *out-of-sample prediction*, and it is generally the most rigorous test of a statistical model.

The ability to predict what is going to happen is one of the main tests of good science generally. This is why replication has such a high value in science. For example, consider research on 'power poses' (see also *The SAGE Quantitative Research Kit*, Volume 7, Chapter 1). Carney et al. (2010) investigated whether assuming certain bodily postures known as 'power poses' could effect a change in the levels of testosterone in their participants' blood. Their study yielded the surprising finding that adopting power poses was associated with raised testosterone levels in a small experimental sample. There was, as far as we know, nothing wrong with the methodology of this study. The methods were all scientifically sound. Given the result from this first study, it was plausible at the time to predict that subsequent studies would also find a relationship between power posing and testosterone. However, when new (and larger) studies examined the same phenomenon, employing much the same methodology as the first study, they found little or no evidence of such a relationship (e.g. Ranehill et al., 2015). This suggests that the initial finding may have been a fluke – a case of bad luck, where a 'statistically significant' association was found in the data, despite there not being any real psycho-physiological process that links body posture to hormone levels, at least not in the way suggested by the original study.[iv]

Let's formalise the logic of out-of-sample prediction for statistical investigations. Out-of-sample prediction refers to the process of testing the predictions from a model developed on one data set by using a different data set. Let's call the data set that we use to develop our model the *development data*, and let's call the data set that we use to test our model the *test data*.

Within the endeavour of science, there may be different ways in which we can investigate a model's power to predict out-of-sample observations. For example, if we conduct a small experiment such as the power poses study, the data from that experiment are our development data. Another study that uses the same research design and experimental procedures, but new participants, will yield test data that we can use to verify or falsify the model developed on the first study. This process is called *replication*.

A related but different procedure is *crossvalidation*. If we conduct a study with a large data set, such as data from a national survey or administrative records, we might incorporate the idea of testing our model on new data by dividing our data set. For example, we might randomly select 50% of our cases to act as the development data set, and the other 50% as the test data set. We carry out all our modelling on the development data, decide on a model that seems best and then investigate how well this model performs on the test sample. This is a simple form of crossvalidation.[v]

Replication and crossvalidation share the idea that our model, developed on one data set, ought to be tested on another. Both can evaluate the stability of our model

---

[iv]The story of the power poses study and its failed replications is a good example of why in general it is advisable to avoid claiming that a theory has been scientifically shown to be correct on the basis of a single study, especially when this single study is a small experiment, even if it is well conducted.

[v]More complex crossvalidation involves repeatedly taking random subsamples from our data set and developing a new model on each random subsample, in each case testing on the remaining data. A thorough treatment of crossvalidation is beyond the scope of this book, but see, for example, Kuhn and Johnson (2013).

across data sets. Yet replication and crossvalidation have slightly different rationales. Replication aims for more than crossvalidation, because it involves data being collected in a different context, such as at a different time and place, as well as potentially by different researchers, potentially using different measurement instruments, and so forth. Replication is thus able to investigate model consistency across contexts, which crossvalidation cannot do.

Even if, at the time that you are modelling your own data set, you cannot perform crossvalidation and cannot yet know whether you will have the resources to perform a replication study in the future, it helps to think that what you are doing at the current moment is to develop a model from a development data set and that in the future your results might be put to the test. This should be a relief and a challenge at the same time:

- A relief, because it means that it is not your responsibility to deliver the ultimate answer to the research question you are posing; instead what you deliver is a plausible model of the data and an empirically justified *contribution* to answering a question.
- A challenge, because you are aware that your result may be put to the test in new data; this should make you cautious and should motivate you to avoid overoptimistic claims based solely on the one data set that you happen to have in front of you.

As eminent statisticians have said, 'Statistics is the science of learning from experience, especially experience that arrives a little bit at a time' (Efron & Tibshirani, 1993, p. 1). In other words, statistics is the science of learning the most we can from the data at hand, while being mindful that our current data set won't necessarily give us the ultimate answer to the questions we are investigating and that in the future new data may come along which might change our conclusions.

Statistical models and uncertainty

To put statistical modelling in the social sciences into perspective, let's contrast it with deterministic modelling. As an example of the development of a deterministic mathematical model, take the experiments conducted by Galileo Galilei (1564–1642), who was interested in how fast objects fall (MacDougal, 2012). In Galileo's time, clocks were not precise enough to measure falling times with sufficient accuracy. Things simply fell too fast for 16th-century clocks. To slow down the process of 'falling', Galileo constructed an inclined plane out of a long piece of wood and made a lengthwise groove in the middle for a bronze ball to roll down on. To reduce friction, the groove was lined with parchment. Galileo generated data by repeatedly rolling a bronze ball down this wooden plane, varying the angle of the plane's slope and meticulously recording the time the ball took to roll a given distance.

After doing this many times, Galileo found a mathematical relationship between the distance that the ball travelled and the time it took to do so: the distance is proportional to the square of the time. In mathematical notation, this can be written as follows: $d \propto t^2$, where d signifies the distance, t signifies the time, and the symbol $\propto$ means 'is proportional to'. This is an example of a deterministic mathematical model. There is no error term, there are no residuals to investigate. Once he had built and set up his equipment, Galileo could repeat his experiments as often as he liked until he was as sure as he could be that his model was able to predict his experimental observations.

Compare this with a typical social science investigation – for example, with Alicia Holland's (2015) study on police operations against street vendors in Latin American capitals (see Chapter 5). It took a long time and much effort to collect the data. To replicate the study, a similar effort would need to be made. Replication in the same cities would need to look at subsequent years, so time needs to go by in order for replication to be possible. Galileo, if one of his experimental runs yielded an unusual result, could say 'Maybe I should renew the parchment on my wooden plane' and start again. In contrast, Holland can't reverse time and set the street vendors, police officers and mayors of Lima, Santiago and Bogotá on their paths again to see how things would turn out in an alternative universe.

At the end of statistical modelling, all we have is estimates, and there always remains uncertainty around these estimates. In such a context, the advantage of using a statistical model, rather than a deterministic one, is in the ability of the statistical model to quantify the uncertainty. This is why, in statistical data analysis, we allow for error terms in our models, why we put margins of error (confidence intervals) around our estimates and why we try to avoid over-optimism when drawing conclusions from our data. A good data analyst is mindful that they would prefer to replicate their study several times to arrive at more precise conclusions. When they are unable to do so immediately, the value of their study lies in providing a piece of evidence that they themselves and others can build on in further research.

So, at the end of statistical modelling, you are under no obligation to have certainty about the right answer to the research question. A statistical model does, however, help you to quantify how uncertain your answer is. In the next section, we will look in some more detail at what it means to be cautious and what it means to be overoptimistic when making claims about findings from a statistical model.

Two ways of getting it wrong: overfitting and underfitting

Two concepts that might help in thinking about statistical models are **overfitting** and **underfitting**. To understand these terms, let's remind ourselves of the aim of a statistical model: to detect the signals contained in the data and to separate them from the noise. Overfitting and underfitting are two ways in which a model can fail in this task. An overfitted model is one that reproduces random noise in the data, rather than detecting the signal. An underfitted model is too simplistic to detect the signal.

A model that is overfitted is usually too large: it contains more predictors, or more transformations of predictors, than is necessary or useful. An overfitted model tends to fit the development data well but does poorly when put to the test of predicting new data. To put the same thing differently, a model is overfitted if its predictions correspond very closely to the development data but fail to predict test data well – while a smaller model exists that might do worse on the development data but does better on the test data.

In practice, overfitting often results from the analyst's desire to arrive at a model that accounts for as much variation as possible in the development data. It can be very seductive to select the model that has the highest R^2-statistic or C-statistic or likelihood, in the belief that this would show that one has chosen the best model. The danger of undetected overfitting is particularly great if the model is never tested on new data.

Contrast this with underfitting. A model that is underfitted is usually too small: it contains too few predictors, or fails to do justice to the true shape of the relationship between the outcome and one or several predictors, for example by ignoring non-linearity. An underfitted model fits neither the development nor the test data well, while a larger model exists that would do better on both the development and the test data.

Underfitting often results from using a statistical model without thought, for example, by putting variables into a default software procedure without thinking about how the model relates to established scientific (sociological, epidemiological, psychological, . . .) theory and knowledge, and without exploring the data, checking assumptions or considering the plausibility of the resulting estimates.

If we are able to replicate our study, or employ crossvalidation, we can check for underfitting and overfitting on the test data. But even if we cannot, it is sometimes possible to identify over- and underfitting with a high degree of plausibility from comparing the model predictions to the observations. I will give a simple example. Figure 6.2 shows estimates of mean annual global temperatures from 1880 through 2018 from NASA's Goddard Institute for Space Studies. The temperatures are being displayed in degrees Celsius relative to a reference period (1951–1980). So a value of 1 on the temperature axis means 'the mean temperature this year was 1°C warmer than the average of 1951–1980'. This is a standard way of displaying such data.

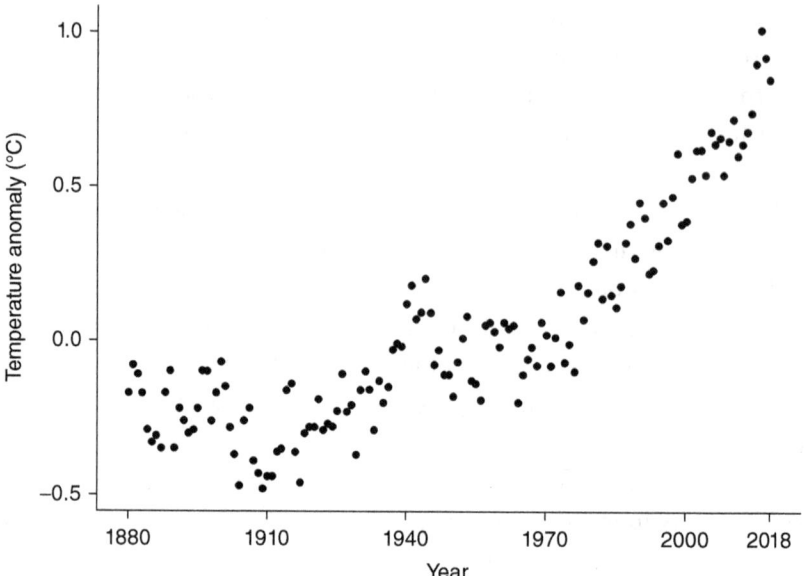

Figure 6.2 Annual global mean temperatures (degrees Celsius), 1880–2018

Note. Temperatures are given as 'anomalies' relative to the reference period (1951–1980). For details, see https://data.giss.nasa.gov/gistemp/. NASA Goddard Institute for Space Studies (GISTEMP Team, 2020; Lenssen et al., 2019).

The most striking feature of this graph is the rapid rise of global mean temperature between 1970 and 2018. There also seems to have been a slight global cooling during the 1940s, which was preceded by a period of mild warming from around 1910 to 1940. The temperatures obviously vary from year to year, but how should we model the relationship between time and temperature? The first step in making that decision is to inform ourselves about how the data were collected and what the data quality is likely to be.

As you might imagine, measuring mean global temperature is complicated for various reasons, which include the following:

- We don't have weather stations on all points of the globe, and those that we do have are not spread evenly.
- Over time, weather stations may change their location or close entirely, and new weather stations may open in other places.
- There are more weather stations today than there were in the past.
- Generally, measurements have become more accurate and frequent over time, due to improvement in measuring techniques.

For all these reasons and more, statistical modelling is required even to produce these estimates of annual global means. So there is some uncertainty (measurement error) in the data, and there is good reason to believe that the uncertainty is larger for older data (which are based on fewer and more uncertain observations) than for data from more recent times (which are based on more, and more precise, observations).

Now consider Figure 6.3. This shows four statistical models that were fitted to the global temperature data. I have called them Model (a), (b), (c) and (d), respectively.

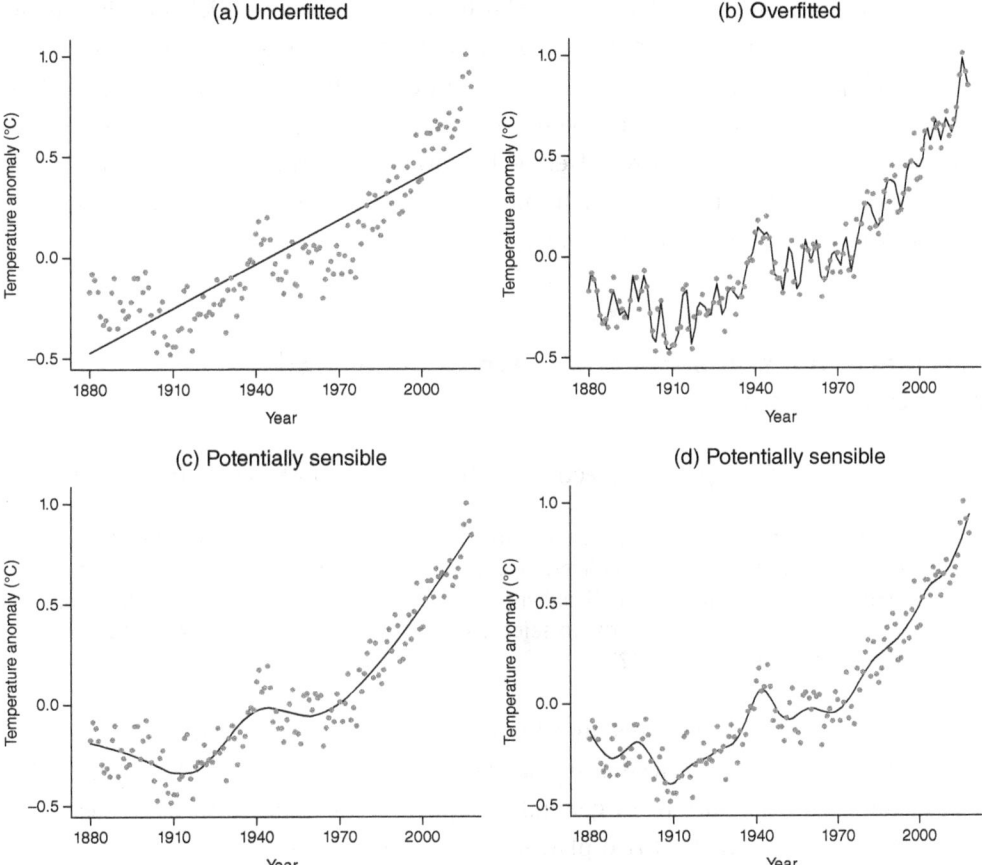

Figure 6.3 Four statistical models fitted to the annual global mean temperature data

Note. Temperatures are given as 'anomalies' relative to the reference period (1951–1980). For details, see https://data.giss.nasa.gov/gistemp/. NASA Goddard Institute for Space Studies (GISTEMP Team, 2020; Lenssen et al., 2019).

Model (a) is underfitted. It represents the trend in global temperatures as a straight line, which clearly does not do the data justice. Residual diagnostics of model (a) would confirm that there is unmodelled non-linearity in the relationship between time and temperature.

In contrast, model (b) is overfitted. Its wriggly line tries to connect almost all dots. In other words, the model produces predictions that aim to almost perfectly reproduce the data. This model achieves almost no simplification. Moreover, since we know that there is measurement error in the data, we know that this overfitted model does not make sense from a scientific point of view. Model (b) is an example of a model that reproduces the noise in the data, rather than detecting the signal.

Models (c) and (d), on the other hand, do seem to make sense. Both bring out the essential features of the data. For example, both suggest a more or less straight-line increase in temperature from around 1970 to 2018. The two models do disagree in some particulars – for example, they differ in their estimates of the size of the short-lived cooling during the 1940s. We could conduct further investigations to explore which of these two models might be more plausible. It might turn out, for example, that model (d) is also overfitted. Detailed knowledge of climate science and other data, especially those that measure human-made emissions of greenhouse gases, are necessary to decide between plausible models.

Is science in a statistical crisis? On p-values and hypothesis tests

You may have read the previous section in this chapter and thought to yourself,

'It may be an interesting pastime to think about subjective decision making in statistics, uncertainty, the desirability of replication, underfitting and overfitting and so forth, but ultimately, isn't it the job of statistical hypothesis tests to provide an objective criterion by which I can decide which model to select? I'll do a likelihood ratio test and that tells me what I should believe, right?'

This section is a response to this question and to this way of thinking.

Indeed, statistical hypothesis tests can be useful in the process of model choice and in model interpretation. In general, the purpose of a statistical hypothesis test is to help us decide which of two plausible models is more consistent with the data. Despite the usefulness of hypothesis tests, at various points in this book I have advised to use a test only after carefully thinking about whether it is necessary, which kind of test is appropriate, and whether other analytic methods might be more meaningful. I have also emphasised that hypothesis tests don't remove uncertainty from a conclusion: if the test yields a small p-value, I have interpreted this as *evidence against the null hypothesis*, or *evidence against the smaller model*, rather than asserting that the alternative hypothesis is 'correct' or that the larger model is 'better'. Conversely, when the p-value is large, I have interpreted this as *little evidence against the null hypothesis*, rather than asserting that 'the null hypothesis is right'. These interpretations recognise that a hypothesis test does not yield a conclusion that is certain. You may find that, in some scientific publications, interpretations of statistical hypothesis tests are somewhat less cautious than this.

Critique of current practice around statistical hypothesis tests

The way hypothesis tests and p-values are currently used by many researchers in social science, psychology, biomedical research and related disciplines has come under critical scrutiny. For many years, statisticians and other scientists have been pointing out that many researchers misunderstand p-values, overuse hypothesis tests at the expense of other statistical techniques, and often report 'statistically significant' results that turn out to be spurious findings (Gigerenzer, 2004; Gigerenzer et al., 1989; Ioannidis, 2005; Simmons et al., 2011). The misuse of p-values and statistical hypothesis tests has many roots and many consequences. One important consequence is that it contributes to the replication crisis in many sciences – whereby a high proportion of published research findings turn out not to be confirmed by attempts to replicate them with new data (Ioannidis, 2005). In recent years, critique of the unthinking use and often incorrect interpretation of p-values, as well as of the routine use of the conventional cut-off of 0.05 to define statistical significance, has intensified (Colquhoun, 2014; Greenland et al., 2016; Wasserstein & Lazar, 2016). Some scholars have argued that we should altogether 'abandon statistical significance' (McShane et al., 2019). Others, although they agree that radical reform of current common practice is needed, wish to retain the concept of statistical significance for some purposes (Ioannidis, 2019).

So what's wrong with p-values? My own position is that nothing is wrong with them in principle. However, p-values are often used in situations where they are not a valid measure of anything, they are often misinterpreted, and sometimes they are intentionally misused to make a conclusion look scientific when it is in fact poorly supported by evidence. I also argue that p-values and 'statistical significance' have inappropriately become the main way in which many researchers, and editors of scientific journals, assess the strength of the evidence for research hypotheses and findings, at the expense of other important considerations, such as the research design, the precision and validity of the measurements used, the quality of the data, and the plausibility of the finding in the light of prior evidence and associated scientific knowledge (Gigerenzer, 2004; McShane et al., 2019).

What is a statistical hypothesis test again?

Let's remind ourselves of the definition of a p-value. To help us do so, we'll consider an example. Imagine that we wish to investigate the sex ratio at birth: the relative frequency with which newborn human children are boys or girls. Suppose that we know nothing about this topic before we start the investigation, and suppose therefore that we start with the hypothesis that a child is equally likely to be born a boy or a girl. Then the null hypothesis of our study will be that the boy to girl ratio of newborn babies is equal to 1. Let's denote the sex ratio at birth by 'SRB'. The null hypothesis states that $SRB = 1$.

A p-value arises from the evaluation of a statistical model. Let's call this model the null hypothesis model, or Model 0. This model encompasses:

- a hypothesis about the world. In our example, we hypothesise that a baby is equally likely to be either a boy or a girl, such that $SRB = 1$.
- assumptions about the process by which data are generated. In our example, we might assume that the numbers of boys and girls each follow a binomial distribution. This would imply, for example, that the sex ratio at birth is the same in all sections of the population. In other words, we assume that $SRB = 1$ not only on average in our population, but also that $SRB = 1$ in all conceivable subgroups of the population.
- assumptions about the procedure the researcher used to conduct the test. For example, a crucial assumption is that the researcher defined the hypothesis and decided on the test procedure independent of seeing the data. We will discuss violations of this assumption presently.
- further assumptions, often implicit, that, if true, would ensure that the data we collect are not biased systematically in one direction or another. In our example of studying the sex ratio at birth, our data set may be derived from birth records, so we need to assume that these birth records are complete and accurate. To give a different example: in surveys, typical assumptions about the data are that we have suitably defined the population of interest, that the sampling frame is representative of this population, that the sample is a true random sample from the sampling frame, and that our measurements are valid and reliable (e.g. that our survey respondents answer questions truthfully and comprehensively).

In summary, once you consider the matter thoroughly, you will see that a null hypothesis model contains a large number of assumptions.

Now, let's imagine that we have collected our data set. Based on the data, we can calculate a test statistic, and from this and our definition of the null hypothesis model (Model 0), we can obtain the probability that Model 0 would have led us to observe this test statistic, or a test statistic further away from the value hypothesised under Model 0. This probability is the p-value.

Suppose that our data set contains 1000 birth records, and suppose we find that 550 children are recorded to be male, while 450 are recorded to be female. So the observed sex ratio at birth is $\widehat{SRB} = \frac{550}{450} = 1.22$. We can use the binomial distribution to find the probability of observing 550 or more children of one sex (from a total of 1000), if Model 0 is true. This probability can be calculated to be $p = 0.0014$. So if Model 0 was true, we would expect to see 550 or more children of one sex about 14 times in 10,000 samples.[vi] This is considered a small p-value in social science.[vii]

If the p-value is small, this means that a reality that accords with Model 0 would be unlikely to produce the data we see. We might interpret this as evidence against Model 0. However, the finding does not tell us which aspect of Model 0 may be false. There are different possibilities. It might be that the null hypothesis is wrong: the sex ratio at birth may not be equal to one. Or it might be that the statistical assumptions don't hold: for example, the proportion of boys may not follow a binomial distribution. Or it might be that one or several of the many assumptions about the data (quality of measurement, accuracy of the records, etc.) have not been met. Of course, on top of all that there is also the risk of Type I error, which occurs when the null hypothesis is correct and all model assumptions are met, but when nonetheless, by bad luck, we have observed a very improbable result and hence have obtained a small p-value.

On the other hand, when the p-value from a statistical hypothesis test is large, this indicates that a reality that conforms with Model 0 would have a high probability of producing the data that we see. We may interpret such a result as lack of evidence against the null hypothesis. Contrary to what many researchers believe, however, a large p-value does *not* in itself constitute evidence in favour of Model 0. This is because a hypothesis test tells us nothing about the relative merits of Model 0 compared to other possible models that make specific predictions.

To illustrate this, imagine that in our sample of 1000 births we had observed 520 boys and 480 girls. Then the two-sided p-value calculated under the assumption of Model 0 would be equal to $p_{\text{Model 0}} = 0.195$. So, if Model 0 was true, we would expect to see 520 or more births in either sex in about one in every five samples. That is quite a likely occurrence, and therefore this result could not be interpreted as strong evidence against Model 0. However, the result does not provide evidence in favour of Model 0 either. This is because our data are also compatible with many other models. For example, suppose a different researcher investigates the same data, but before seeing the data they have hypothesised that the number of newborn boys exceeds the

---

[vi]This is a two-sided p-value; that is, it's the probability to observe either 550 or more boys, or 550 or more girls, under the assumption of the null hypothesis that $SRB = 1$.

[vii]What is considered a small p-value depends on convention and the area of science we are working in. For example, in particle physics, an experimental effect is considered robust if the observed result is at least 5 standard errors away from the mean expected under the null hypothesis. This is called the 'five sigma rule' and corresponds to a significance level of $p < 0.000000287$ (one-sided test). That is, the null hypothesis is rejected only if, under the null hypothesis, the observed result and results further away from the null hypothesis value have a probability smaller than about 3 in 10 million (Lamb, 2012).

number of newborn girls by about 5%. We might call this Model 1: $SRB = 1.05$.[viii] Our data (520 boys and 480 girls) yield a large p-value for this hypothesis also: $p_{\text{Model1}} \approx 0.6$. In fact, there are many plausible models that our data would be compatible with according to the p-value criterion, had we hypothesised these models in advance. Thus, a large p-value in itself is no evidence at all for the null hypothesis.

In social science, countless hypothesis tests have been conducted to assess the evidence for a relationship between two variables, X and Y. The null hypothesis almost always is that X and Y are uncorrelated. Suppose that a statistical test of this hypothesis yields $p = 0.195$. It is unfortunately not unusual to find this result reported as '$p = 0.195$, so there is no relationship between X and Y'. As we have seen, this conclusion does not follow.

Misuses and misunderstandings of p-values

In the context of multivariate statistical modelling, many authors have pointed out some stark misuses of hypothesis tests that are part of current scientific practice.

P-value hacking

Imagine a researcher who has in front of her a large data set with many variables but no or little theory that would predict which variables might be related to which others. Imagine further that this researcher knows that it would benefit her academic career if she was able to publish a novel 'finding' in a scientific journal. One way for the researcher to produce such a 'finding' is the following: She explores all possible relationships between the variables in her data set and uses a statistical hypothesis test each time. Even if none of the variables are truly related to one another, by using the $p < 0.05$ threshold for a 'statistically significant' result, she could expect about 1 in 20 of the statistical hypothesis tests she carries out to yield an apparent 'statistically significant' result. So if our researcher conducts many statistical tests, looking, for example, at relationships between different pairs of variables, the chance is high that at least one of the explored relationships will seem to be 'statistically significant' in this way.

Now, the researcher has the opportunity to cheat: she can write her research paper reporting only the 'statistically significant' results, ignoring the analyses that yielded large p-values. She then writes an introduction to her research report that makes it

---

[viii]Current scientific knowledge regards 1.05 as the best estimate of the human sex ratio at birth (males to females) for most of human history (Chao et al., 2019). See also the section 'Exclusive Focus on Hypothesis Tests Distracts From Other Useful Purposes of Models'.

look as though she had hypothesised the 'significant' results before seeing the data. Thus, unsuspecting readers would be led to believe that they are reading a report by a scientist who has done careful research and whose results bear out her carefully considered hypotheses. The low p-value attached to the reported finding gives a widely recognised stamp of scientific objectivity. So it looks as though the researcher's theory is well supported by the data. But, in truth, the p-value is meaningless, because the researcher looked for a small p-value to tell her which hypothesis to write about, rather than using data to put a hypothesis to a rigorous test.

The researcher's trick is somewhat akin to that employed by a marksman who aims to show that he can hit his target with every shot of his gun (Mayo, 2018). The trick works as follows: our marksman fires his gun at a large barn door without aiming too precisely. Then he goes and draws a target around each bullet hole in the barn door. In this way, every shot can be seen to have hit the target. To an observer who sees how it is done, it is obvious that this procedure says nothing about our marksman's skill at shooting. However, an observer who does not see the procedure, but only the result, might take this as evidence in favour of our marksman's claim to never fail his target. In the same way, a low p-value achieved through the trick of multiple testing constitutes no evidence for any scientific hypothesis – but if the procedure of multiple testing is hidden from view, then it might seem to do so.

The procedure of obtaining small p-values via undeclared multiple testing is called 'p-value hacking'. It's also sometimes referred to as 'fishing for significant results', or 'hypothesising after results are known'. The result of the trick is that our researcher has a publication to her name, which may enhance her reputation and career. But the published finding is worthless as a contribution to knowledge, and may even be harmful to scientific progress. Other researchers may spend time and grant money on efforts trying to replicate the p-hacked result, or may even assume it to be true and build further investigations and research programmes on it, which are then likely to be doomed to failure.

The journalist John Bohannon (2015) intentionally used p-hacking to expose the ease with which poor research can achieve high visibility in the media. He and his colleagues conducted a small study about nutrition. They randomised 16 participants into three groups. The first group was instructed to follow a low-carb diet for 3 weeks, the second was asked to follow the same diet with the addition of dark chocolate, and the third was told to make no change to their usual diet. The study measured 18 different outcome variables. Bohannon and his colleagues did not know in advance what they would find, but they knew that their study had a high chance of yielding *something* 'statistically significant'. As Bohannon says, 'With our 18 measurements, we had a 60% chance of getting some "significant" result with $p < 0.05$. (The measurements weren't independent, so it could be even higher.) The game was stacked in our favor' (Bohannon, 2015).

As it turned out, the 'dark chocolate' group had the fastest weight loss, and the weight difference to the other groups was 'statistically significant'. Bohannon (2015) and his co-conspirators then published the results, accompanied by a press release claiming that eating chocolate leads to weight loss. This was widely and uncritically reported in newspapers and online publications across many countries before Bohannon exposed his finding as a spoof. His account of this hoax is highly recommended reading (Bohannon, 2015).

Researcher degrees of freedom: the garden of forking paths

The effect of p-value hacking can also be achieved by researchers who do not set out to deceive and who genuinely aim to do their best, but who misunderstand statistical hypothesis tests (Gelman & Loken, 2014; Simmons et al., 2011). For example, suppose you have theoretical reasons to suspect that X is related to Y and have collected data to investigate this. In the first model you fit to your data, you find that, to your surprise, the relationship between X and Y is not statistically significant ($p > 0.05$, say, for the null hypothesis of 'no relationship between X and Y'). You wonder why the data do not conform to your prediction and surmise that possibly X is only related to Y in a subgroup of the data (say, only in women). You try that analysis. Again, the relationship has a p-value higher than 0.05. You realise that the highly reputed journal, where you had hoped to publish your research, would likely not accept that your data support your theory, since you cannot claim 'statistical significance'.

You then consider that maybe you should have excluded some records from the data set. You investigate subgroups and notice that the relationship between X and Y looks stronger in younger than in older people. Could it be that the relationship between X and Y does not hold for older people? You exclude everyone over 70 years from your analysis and run your model again. Once more, among women, the relationship between X and Y is 'not significant', and neither is it in the whole sample. However, looking at the coefficients, it occurs to you that the results suggest that the relationship between X and Y is actually stronger in men than in women in your data set. So you run the model a fourth time, this time including in the sample only men under 70 years. Lo and behold, this time the p-value is smaller than 0.05. So you believe that you have finally found evidence for your theory. Suddenly it all makes sense!

Inspired by your novel finding, you identify a theoretical reason why the relationship between X and Y should be stronger in men than in women and why it should only exist in people under 70 years. You report the result, dutifully saying that you tested the male, female and whole sample, and also reporting that you excluded people over 70 years. However, your report does not necessarily make clear to the reader that you have investigated various subgroups of your data and tested the relationship between X and Y five times before you found your result, and that you identified your hypothesis about the right subgroup to investigate after seeing the data. The p-value you report looks scientific, but it is actually meaningless, because the procedure you used to arrive at your model had a high chance of finding some apparently 'statistically significant' result, even if in reality there is absolutely no relationship between X and Y. This is because, had you found a 'statistically significant' result in your first try, you would have stopped there and reported this. On the other hand, if even the fifth try had not yielded a small p-value, you might have conducted further investigations, until the desired 'statistical significance' had been attained.

In other words, you have behaved like the marksman who draws his targets around his bullet holes, but without realising that you did so. Put more technically, given the data analysis procedure, the probability of finding a 'statistically significant' association between X and Y in *at least one* analysis model, under the null hypothesis that X and Y are unrelated, was not 5% (as it would be for a valid statistical test using a 5% cut-off for rejecting the null hypothesis), but considerably higher. In fact, if we are prepared to try sufficiently many analyses, then the probability of finding some p-value to be smaller than 0.05 might approach the value 1.

Gelman and Loken (2013) have evoked the image of a maze, a 'garden of forking paths', to explain how researchers may deceive themselves into thinking that they are conducting valid statistical hypothesis tests when in fact they are not. At various points in the data analysis, the researcher makes a decision regarding how to analyse the data. Once one decision is made (one path is taken), another decision presents itself. Each decision in itself may feel entirely justified to the researchers, based on the data they see. However, the point is that each decision depends on the particular data set at hand and is influenced by the hunt for a 'statistically significant' result.

Imagine now if we were able to collect a new data set from the same population. Then, simply due to sampling variation, we would get different data, exhibiting different apparent patterns. Following the 'forking paths' procedure, the researchers would again make decision after decision, until they eventually arrive at a 'statistically significant' result. This time, maybe, the relationship between X and Y seems to hold only among women under 40 years (instead of in men under 70 years). In another analysis, we might find apparent evidence for the relationship among both men and women, but only if we exclude the unemployed, define X in a certain way, exclude certain outliers or . . . . Thus, without the intention to deceive, researchers may in fact deceive their readers *and* themselves by trying out different analyses until they find a result that is 'statistically significant'. This is the 'garden of forking paths'. This phenomenon has also been described under the name *researcher degrees of freedom* in a seminal paper by Simmons et al. (2011).

To summarise, as Gelman and Hennig (2017) put it succinctly, 'researchers often rely on the seeming objectivity of the $p < 0.05$ criterion without realizing that [the] theory behind the p-value is invalidated when analysis is contingent on data' (p. 969). p-Values are only indicators of the evidence against a null hypothesis if the model under which they were calculated was specified before the analyst saw the data. This is not to say that exploratory analyses and the use of data to build a model should not be undertaken. Of course they should. But p-values are meaningless as evaluations of the findings from such analyses.

Exclusive focus on hypothesis tests distracts from other useful purposes of models

Even when p-values are calculated under valid conditions and interpreted correctly, a one-sided focus on p-values can unduly limit scientific practice. To illustrate this, let's look at what's usually regarded to be the first statistical hypothesis test in history.

John Arbuthnot[ix] (1667–1735) was a Scottish physician and polymath. He observed that men were more likely than women to die before they had a chance to have children. Men were more likely to die young due to being at higher risk of accidents than women, and at higher risk of being drafted as a soldier into a war. But, Arbuthnot argued, should not God[x] have arranged it so that there is one man for every woman? Then, God should arrange it so that more male than female babies are born. To investigate his hypothesis, Arbuthnot looked at records of births over 82 years, from 1629 to 1710. Table 6.1 gives an illustrative extract of his data.

Table 6.1 An extract from the data on births in London used by Arbuthnot

Year	Males	Females	Ratio
1629	5218	4683	1.114
1630	4858	4457	1.090
1631	4422	4102	1.078
⋮	⋮	⋮	⋮
1708	8239	7623	1.081
1709	7840	7380	1.062
1710	7640	7288	1.048

Note. Data from Arbuthnot (1710) and Friendly et al. (2020).

[ix]My use of this story to illustrate the limitations of statistical hypothesis tests owes a debt to Gigerenzer and Marewski (2014). Steven Stigler (2016) also gives a thorough and lucid account.

[x]It was a common belief among Christian intellectuals in Arbuthnot's time that science was a way to understand the creations of God.

Arbuthnot (1710) found that in each year the number of males born was larger than the number of females. He then calculated the probability that this would occur by chance in 82 consecutive years, under the assumption that a baby is equally likely to be born male or female. This probability is

$$p = \left(\frac{1}{2}\right)^{82} = 2.07 \times 10^{-25} = 0.0000000000000000000000000207$$

So the probability of observing more male than female births in all of 82 years is vanishingly small, under the assumption that every baby has an equal chance to be born female or male. Therefore, Arbuthnot (1710) concluded, we have evidence that 'provident Nature, by the disposal of its wise Creator, brings forth more Males than Females' (p. 188). Or, as some scientists would put it today, '$p < 0.05$, therefore we have evidence for the effect of divine providence on the sex ratio at birth'. Do we?

Let's look at Arbuthnot's procedure from a modelling perspective. Let's call the male–female ratio of births SRB (sex ratio at birth – see the section 'What Is a Statistical Hypothesis Test Again?'). A suitable model for the data might be

$$\widehat{SRB} = SRB + \varepsilon$$

where $\widehat{SRB}$ is the observed sex ratio at birth, SRB is the true ratio and ε is the difference between the two (the statistical 'error').

The null hypothesis model of Arbuthnot's test, Model 0, assumes that $SRB = 1$, so that the observed ratio of male to female births in any given year is $\widehat{SRB} = 1 + \varepsilon$.

Arbuthnot's alternative hypothesis, Model 1, assumes that $R > 1$ – that is, that overall more babies are boys than girls.

Arbuthnot's test of Model 0 yields a low p-value, as we saw above. This means that the data are very unlikely to have been generated by this model. The evidence suggests that $SRB > 1$. So Arbuthnot's logic in rejecting Model 0 and accepting Model 1 is exactly the logic of statistical hypothesis testing.

But Arbuthnot's argument also reveals some of the limitations of hypothesis testing (Gigerenzer & Marewski, 2014). Specifically:

- Arbuthnot's p-value of 2.07×10^{-25} is the probability of observing the data given that Model 0 is true. Arbuthnot's accepted statistical model ($SRB > 1$) is plausible given the data, but this does not in itself indicate that his substantial theory, divine intervention, provides the correct explanation for his statistical observation. Crucially, Arbuthnot did not use his hypothesis to predict what the sex ratio at birth should be if his theory was right. His adopted Model 1, $SRB > 1$, is rather unspecific. So Arbuthnot has not really put his theory to the test. His statistical hypothesis test rejected a particular model, but this does not allow him to claim evidence for his theory.

- By concentrating on the rejection of his null hypothesis, Arbuthnot failed to make a perhaps more interesting use of his data – namely to estimate the male–female birth ratio. This might have been a more useful scientific result.[xi]

We may find similar limitations in some scientific publications of the current age, where statistically significant findings are claimed as evidence for substantive theories, on the basis of having rejected some null hypothesis (usually a null hypothesis of 'no effect'), without further evidence that would support the authors' favoured explanation of the effect they claim to have identified. I hasten to add that this criticism does not apply to the studies I have used as examples in this book, each of which contains a substantial and well-argued section on the theory that inspired and informed the investigation, as well as thorough methodological reflections regarding the opportunities and limitations of the data.

In conclusion, I would argue that Arbuthnot's use of a statistical hypothesis test is fine as far as it goes. His finding that the sex ratio at birth is not equal to 1 is important, even if we don't accept his explanation for it. However, I believe that statistical analysis would serve science better if we

- use theories to make specific predictions about what we expect to find in the data (before seeing the data) and then use data to test those theories, and
- use data to estimate the size of quantities of interest, rather than just reject some null hypothesis about them (e.g. to focus on estimating the size of a regression parameter, rather than just assessing the null hypothesis that it is zero).

Beyond this book: other types of models

After you have read this book, one thing that might happen is this: you may find yourself with a question that you would love to research and a data set that you can analyse to address this question. And then it turns out that one of the models introduced in this book is suitable for your data. Maybe your outcome is dichotomous, and you can use logistic regression. Or your outcome is a count variable, and careful modelling reveals that a zero-inflated Poisson model fits the data well. How lucky! All the hard work reading through that modelling book was actually useful!

[xi]Since Arbuthnot's time, much scientific work has been done to estimate the sex ratio at birth. There is broad agreement that the 'natural' sex ratio at birth, for most of human history, is about 1.05 (male births divided by female births), although there may be some regional and ethnic variation. This is an important scientific result. For example, this 'natural' ratio has been used as a reference value to assess the prevalence of child sex selection through abortion. There is strong evidence that the birth sex ratio has become larger in some countries since the 1970s (Chao et al., 2019).

However, another situation may also arise: you may find yourself with a question that you would love to research and a data set that you can analyse to address the question. But none of the models this book introduces seems appropriate. Maybe the count outcome is so overdispersed that even a negative binomial regression won't account for the excess variance. Or some of the relationships between outcome and predictors are non-linear in ways that cannot be reconciled with any transformation that you know. Or you might be faced with a data set that violates the assumption of independence of observations. In such cases, even though this book by itself may not provide the answer, it is certainly one of the aims of this book to act as a stepping stone towards you learning about other, possibly more advanced, modelling techniques.

Here are some situations that might require you to go beyond the methods discussed in this book when modelling categorical and count outcomes:

- *Dependency in data:* Some social science data have a clustered data structure (also known as a dependent or hierarchical data structure). Examples include data on children clustered in schools, patients clustered in hospitals or voters clustered in areas. Such data violate the assumption of independence (see Chapter 1). Several types of models can be used to analyse clustered data, including *generalised estimating equations* and *multilevel models*. A good introduction to multilevel models for categorical and count data is the book by Rabe-Hesketh and Skrondal (2012).
- *Non-linear relationships:* Non-linear relationships between predictors and outcomes can sometimes be modelled by transforming the predictors, as we have seen in several examples throughout this book. However, finding a suitable transformation is not always easy. To make the process easier, software algorithms have been developed that try a range of potential predictor transformations and identify the set of transformations that fits the data best. These algorithms use *fractional polynomials* (Royston & Sauerbrei, 2005; Sauerbrei & Royston, 1999). When the shape of the relationship between predictor and outcome is so complicated that no transformation can satisfactorily model it within a generalised linear model, *generalised additive models* (GAM) might instead be used. GAMs account for non-linear relationships by fitting a smoothed curve to the data. Models (b), (c) and (d) in Figure 6.3 have been produced by GAMs. A good place to learn about GAMs is the book by Simon Wood (2006).
- *Violation of assumptions:* For some data sets and research questions, it may not be possible to find a statistical model that meets all its assumptions – or, at least, we might not be certain that the assumptions underlying our model are justified. *Robust standard errors* can help to draw valid inferences in such cases, or in other words robust standard errors might allow us to calculate valid confidence intervals and valid statistical hypothesis tests about model coefficients even if our model fails to meet one of its assumptions. Sandwich estimators and the bootstrap are two types of robust standard errors that you might consider in such a situation. A good introduction to the bootstrap is given by Efron and Tibshirani (1993). Commands to calculate robust standard errors for regression coefficients are available in good statistical software packages.

When you analyse categorical and count data for your own research, there are many more kinds of tricky situations that you might encounter, and more paths you may want to explore before deciding on an appropriate model for your data. If you have made it to the end of this book, you should in such a situation be well equipped to ask the right questions, seek advice or consult other sources as necessary and make your own decisions. Good luck!

GLOSSARY

Binary variable: See *Dichotomous variable.*

Categorical variable: A categorical variable is one whose values are categories, rather than numbers. The categories may be ordered or unordered. For example, *country of birth* is an unordered categorical variable, while *highest qualification* is ordered. Unordered categorical variables are also often referred to as nominal variables, while ordered categorical variables are often called ordinal variables. A variable that has precisely two categories is called dichotomous or binary.

Centring: Centring is a type of variable transformation that defines the transformed variable to be equal to the original variable minus a constant number c. That is, $trans(X) = X - c$. This is called centring X on c. This means that $trans(X) = 0$ whenever $X = c$, and that $trans(X)$ records the positive and negative deviations of the values of X from c. Typically, variables are centred around their mean or their median, or a rounded number close to the mean or median. In the context of regression models, centring of predictors sometimes has the advantage of making the intercept of the model meaningful in cases where X never takes the value 0. Another purpose of centring is to reduce collinearity when using both a predictor and its square transformation in a model. If X is a variable that has only positive values (such as age), and I wish to use both X and X^2 as predictors, then I will find that X and X^2 are highly correlated. If I instead mean-centre X and use as my predictors $trans_1(X) = X - \bar{X}$ and $trans_2(X) = (X - \bar{X})^2$, the collinearity will be much reduced.

Coefficient: A coefficient is a number by which a variable or a constant is multiplied within an equation. For example, in the fitted statistical model equation, $\hat{Y}_i = 0.5 + 3X_i$, the numbers 0.5 and 3 are coefficients. In this case, the coefficient 0.5 represents the intercept and 3 represents the slope. The intercept is called a constant coefficient, since it is not multiplied with a variable. (You can think of it as being multiplied by 1.) In statistical modelling, we use data to estimate the coefficients of a statistical model. The true coefficients, which we usually do not know, are called parameters. Our aim in statistical analysis is to estimate these parameters. A typical table showing the results of a statistical model reports coefficient estimates – that is, estimates of the intercept and slopes of our statistical model. The term *coefficient* is also used more generally to indicate a statistical measure, as in 'correlation coefficient', 'Gini coefficient', 'coefficient of determination', and so on.

Coefficient of determination: See *R-squared*.

Confounder, confounding: A confounder is a variable that is related to both the predictor and the outcome, and thus causes an association between the predictor and the outcome. An association between a predictor and an outcome is called 'spurious' if it is entirely caused by one or several confounders, while there is no causal relationship between the predictor and the outcome.

Continuous variable: A continuous variable is a numeric variable that can take any value within its possible range. For example, age is a continuous variable: a person can be 28 years old, 28.4 years old or even 28.397853 years old. Age changes every day, every minute, every second, every millisecond, so our measurement of age is limited only by how precise we can or wish to be. Contrast this with a discrete variable, which can only take particular values within its range (e.g. 1, 2, 3, . . .). In practice, continuous variables such as age are often measured with a moderate degree of precision that makes them look like discrete variables in a data set. For example, age might be measured in whole years, but conceptually we still think of it as continuous.

Count variable: A count variable is a discrete numerical variable whose values represent counts. Some examples are number of accidents, number of police operations, number of children. Count variables take as their values the non-negative integers: 0, 1, 2, 3, . . . To be comparable across units of analysis, count variables are often defined relative to a common reference in time or space or population, such as number of accidents in 1 year, number of police operations in a district in 1 year, number of children born to one woman.

Deviance residual: A deviance residual is a type of standardised residual that is used for assessing model assumptions in a fitted generalised linear model.

Dichotomous variable: A dichotomous variable is one that has exactly two values. A synonym for dichotomous is binary.

Discrete variable: A discrete variable is a numeric variable that can only take particular, 'discrete' values. For example, count variables are discrete variables; they can take the values 0, 1, 2, 3, and so on. Number of children is a discrete variable: you can have zero children, one child, or seven children, but not 1.5 children. This is in contrast to a continuous variable, which can take any value within its range.

Dummy variable: A dummy variable is a binary variable whose only two possible values are 0 and 1. Dummy variables are used as a device for representing categorical predictors in a statistical model. Take the variable "smoking status" with three categories: 'never smoked', 'ex-smoker', 'current smoker'. This might be represented by two dummy variables: one dummy identifies current smokers by having the value 1 for current smokers and the value 0 for all others; a second dummy identifies ex-smokers in the analogous way. The remaining category, 'never smoked', is called the reference category. In general, to represent a categorical predictor with k categories, we need $k - 1$ dummy variables. Dummy variables are also called indicator variables in the statistical literature.

Effect, effect size: In statistical modelling, the word 'effect' is used to denote the relationship between a predictor, X, and an outcome, Y. This terminology sounds as if it implies a causal relationship, whereby X 'has an effect' on Y. However, the word effect is also sometimes used loosely in situations where a causal relationship cannot be demonstrated, such as in the analysis of cross-sectional surveys. The size of the effect, in either sense of the word, can be estimated by the slope coefficient of X in a model predicting Y. In some contexts, it is useful to transform the slope coefficient to facilitate interpretation of the size of the effect. Thus, in linear regression, sometimes standardised coefficients are used. In binary and ordinal logistic regression, the customary measure of the effect size is the odds ratio. In multinomial logistic regression, it is the relative risk ratio. Effects of predictors on count outcomes are often expressed as rate ratios.

Equidispersion: A count variable is said to be equidispersed when its mean is equal to its variance. Equidisperson is one of the properties of the Poisson distribution. Therefore, equidispersion, conditional on the values of the predictors, is an assumption made by Poisson regression.

Error: The term *error* has a specific meaning in statistics. The error is the difference between a prediction from a true model and a true value. For example, consider the following statistical model:

$$Y_i = \beta_0 + \beta_1 X_i + \varepsilon_i$$
$$\varepsilon_i \sim NID\left(0, \ \sigma^2\right)$$

where ε_i are the errors and are equal to $\varepsilon_i = Y_i - \left(\beta_0 + \beta_1 X_i\right) = Y_i - \hat{Y}_i$.

All statistical models discussed in this book make assumptions about the distribution of the errors. For example, Poisson regression assumes that the errors are independent and that their variance is equal to the predicted mean. Errors should be distinguished from *residuals*. Residuals can be considered estimates of the true errors.

Estimate: In statistics, an estimate is a number which has been calculated from data and is considered a measurement of a parameter. For example, the mean height calculated from a random sample of 1000 people can be considered as an estimate of the mean height in the population from which the sample was drawn. Similarly, the slope coefficient of the regression of child's height on parents' average height, calculated by statistical software from data about 1000 families, is an estimate of the slope coefficient in the population.

In mathematical notation, estimates are usually marked with a 'hat': for example, $\hat{\beta}$ (pronounced: 'beta-hat') denotes an estimate of β.

Estimating a model. See *Fitting a model*.

Excess zeroes: Some count variables have more zero counts than would be expected under a theoretical distribution, such as the Poisson or the negative binomial distribution. These extra zero counts are called excess zeroes. Zero-inflated and hurdle models are techniques for modelling a count outcome with excess zeroes.

Exponentiation: Exponentiation is a mathematical operation involving two numbers, the base b and the exponent n. The exponent is also sometimes called the power. Exponentiation is written as b^n, and indicates that the base is to be multiplied by itself n times. For example, if we wish to calculate $2 \times 2 \times 2$, this is called exponentiating the base 2 to the power 3 and can be written succinctly as 2^3. The result is, of course, 8. When the base is not specified, it is usually assumed to be Euler's number, which is $e = 2.71828 \ldots$ The phrases 'exponentiating x' or 'taking the exponent of x' are often used as a shorthand for calculating e^x, and this is often written as $\exp(x)$. The latter notation is particularly convenient in cases where long mathematical terms are to be exponentiated: it is often easier to write and read $\exp(\beta_0 + \beta_1 X_1 + \beta_2 X_2)$ than $e^{\beta_0 + \beta_1 X_1 + \beta_2 X_2}$.

Fitting a model: Fitting a model to a data set means using the data to estimate the model parameters. This is also called 'estimating a model'. For example, Francis Galton's model of the relationship between parents' heights and adult children's heights is *Child's height* $= \alpha + \beta \times$ *Parents' height*. He fitted this model to data on 928 child–parent pairs, and thereby obtained estimates of the parameters α and β.

Generalised linear model: The generalised linear model (GLM) is a generalisation of linear regression to accommodate outcomes with non-normal error distributions. In a GLM, the outcome is related to a linear combination of a set of predictors via a link function. The link function is a mathematical transformation of the outcome. GLMs allow error distributions other than the normal distribution. Linear, logistic, Poisson and some types of negative binomial regression are examples of GLMs. The zero-inflated and hurdle models introduced in Chapter 5 are not GLMs.

Goodness of fit: The goodness of fit of a statistical model describes how close the observations in a data set are to the values predicted by the model. Various measures and techniques are used to assess goodness of fit, such as graphical exploration of residuals, the R^2-statistic in linear regression, and calibration plots and the Hosmer–Lemeshow test in logistic regression.

Heteroscedasticity: Heteroscedasticity means 'differences between variances'. The term is used to describe a situation where the variance of a variable differs across different groups in the data set. In regression modelling, we often consider how the variance of the errors (or the variance of the observed values) depends on the predicted mean. For example, in a Poisson regression, we expect heteroscedasticity because we assume that the variance of the observed counts is equal to the predicted count.

Homoscedasticity: Homoscedasticity means 'equality of variances'. The term is used to describe a situation where the variance of a variable is the same across different groups in the data set. In regression modelling, we often consider whether the variance of the errors (or the variance of the observed values) depends on the predicted mean. For example, homoscedasticity is one of the assumptions of linear regression; that is, for statistical inference about linear regression coefficients to be valid, we must assume that the error variance is the same, regardless of the predicted mean.

Incidence rate ratio: The incidence rate ratio (also: rate ratio) is a common measure of the effect of a predictor on a count outcome in Poisson and negative binomial regression models. It represents the ratio of the predicted counts of two observations whose values on the relevant predictor differ by one.

Inference, inferential statistics: Inferential statistics is the art of using a sample to draw conclusions about a population, or using a data set to draw conclusions about a process. Tools of inferential statistics used in this book are confidence intervals about coefficients, and statistical hypothesis tests for model comparison and investigating hypotheses about model parameters.

Intercept: The intercept of a regression equation is the predicted value of the outcome variable when all predictors are equal to zero.

Likelihood: The likelihood is the probability of the data given the coefficient estimates of a fitted model. In many types of statistical models, the coefficient estimates are found via maximising the likelihood (see maximum likelihood estimation). The logarithm of the likelihood, the log likelihood, is used in the calculation of the likelihood ratio test statistic, and it is also part of the formula for calculating information criteria.

Likelihood ratio test: The likelihood ratio test is a statistical hypothesis test that compares two nested statistical models with one another. In a typical application of the test, the larger model features some predictors that are not part of the smaller model. The null hypothesis of the test states that the smaller model provides the same predictive power as the larger model. The alternative hypothesis states that the larger model predicts the outcome better than the smaller one.

Log odds: The log odds are the logarithm of the odds of an event: $\log \text{odds} = \log\left(\frac{p}{1-p}\right)$. They result from the logit transformation, which is the transformation of the outcome probability used in binary and ordinal logistic regression.

Logarithm: A logarithm is the answer to the question: b to the power of what gives n? For example, if $b = 2$ and $n = 8$, the question is 2 to the power of what gives 8? The answer is 3, since $2^3 = 2 \times 2 \times 2 = 8$. This is written as $\log_2 (8) = 3$. The number b is called the base of the logarithm, so in this example, the base is 2. Of particular importance in mathematics and statistics is the logarithm to base e ≈ 2.71828..., also called Euler's number. A logarithm to the base e is called the natural logarithm. If the base is not specified, such as when writing log(4), then it is usually assumed that the natural logarithm is meant. This is the case in this book. In statistical modelling, logarithms are used in variable transformations. Log transformation of predictor variables are sometimes used to deal with non-linear relationships. Log transformations of the outcome are often used in Poisson and negative binomial regression, for example. Logistic regression involves log transforming the odds of the outcome event.

Logit transformation: The logit transformation, applied to a probability p, is $\text{logit}(p) = \log\left(\frac{p}{1-p}\right)$. This calculates the log odds (logarithm of the odds). The logit transformation is used to transform the outcome event probability in binary and ordinal logistic regression models.

Maximum likelihood estimation: Maximum likelihood estimation is a method for finding the 'best' coefficient estimates for a statistical model, given a set of observed data. It proceeds by maximising the likelihood, which means, out of all possible coefficients, we choose as our estimates those that have the highest probability of producing the data we have observed. Maximum likelihood estimation involves mathematics that is beyond the scope of this book. In practice, most researchers rely on statistical software to apply maximum likelihood and find the best coefficient estimates.

Nominal variable: A nominal variable is a categorical variable whose categories cannot be ranked or ordered. Examples of nominal variables are country of birth, ethnicity and choice of dish at a restaurant.

Numeric variable: A numeric variable is one whose values are numbers that represent meaningful measurements of some quantity. Numeric variables may be continuous or discrete.

Odds: The odds of an event are defined as the probability of the event occurring, divided by the probability of the event not occurring. If we define p to be the probability of the event occurring, then $odds = \dfrac{p}{1-p}$. For example, if the probability of a patient surviving a surgical operation is 0.9, then the odds of survival are $^{0.9}\!/_{0.1} = 9$. This is sometimes expressed as 'the odds are 9:1 in favour of the patient surviving'.

Odds ratio: The odds ratio is a measure of effect size in the context of binary and ordinal logistic regression. It represents the ratio of the odds in two cases with different characteristics. Consider the odds of having a disease. Suppose that for an 80-year-old patient these odds are equal to 0.1, and for a 60-year-old patient they are equal to 0.05. Then the odds ratio (OR) associated with this 20-year age difference is $OR = \dfrac{0.1}{0.05} = 2$. That is, the 80-year-old has twice the odds of disease compared to a 60-year-old.

Ordinal variable: An ordinal variable is a categorical variable whose categories have a natural order or ranking. An equivalent term is ordered variable. An example of an ordinal variable is 'highest qualification', which might have the categories: 'no qualification', 'completed secondary school', 'high school/A-level', 'university degree or equivalent'.

Outcome, outcome variable: In this book, I use the term *outcome* or *outcome variable* to denote the variable that a statistical model aims to predict. Synonymous terms in this context are dependent variable, or response.

Overdispersion: A count variable is said to be overdispersed when its variance is larger than its mean. A statistical distribution that can model an overdispersed outcome variable is the negative binomial distribution.

Overfitting: Overfitting is the practice of optimising the fit of a model to a current data set, without taking into account the model's likely suitability for new data sets. An overfitted model is one that has been optimised to fit very well to a particular data set, but that does poorly when applied to other data sets. Some of the variation that an overfitted model appears to 'explain' is, in fact, random variation unique to the data set on which the model has been fitted, and unrelated to any effects of predictor variables on the outcome. This is why an overfitted model fails to predict what will happen when new data are collected.

Parameter: A parameter is a quantifiable property of a population, or process. Examples of parameters are the mean height of all adult residents in Greenland, and the average effect of a medicine on human blood pressure. The value of a parameter tells us something interesting or important about the population or process – for example, whether a medicine tends to reduce or increase blood pressure, and by how much. In general, parameters are unknown, but they can be estimated from data. For example, a representative sample of Greenland residents might allow us to estimate the parameter 'average height of all Greenlanders'. A randomised placebo controlled trial might allow us to estimate the parameter 'average effect of a medicine on blood pressure over and above the placebo effect'. Parameters also feature in statistical models. For example, the model

$$Y_i = \beta_0 + \beta_1 X_i + \varepsilon_i$$
$$\varepsilon_i \sim NID\left(0,\ \sigma^2\right)$$

contains the parameters β_0, β_1 and σ^2. These can be estimated from a data set that contains sample measurements of the variables X and Y.

Predicted value: The predicted value of a fitted regression model is the outcome value that, according to our model estimates, is most likely given the values of the predictors.

Predictor, predictor variable: In this book, I use the term *predictor* or *predictor variable* for a variable that is used in a statistical model to predict the outcome. Synonymous terms are independent variable, explanatory variable and exposure. The terms *covariate* (for continuous predictors) and *factor* (for categorical predictors) are also sometimes used.

Rate ratio: See *Incidence rate ratio*.

Relative risk: In the context of multinomial logistic regression, a relative risk is the ratio of the probability of being in one outcome category over the probability of being in a specific other outcome category. Multinomial logistic regression models the log relative risk of two outcome categories. The term *relative risk* has a different meaning in other contexts. In epidemiology, it denotes the ratio of the risk of disease between two groups.

Relative risk ratio: The relative risk ratio (*RRR*) is a measure of the effect of a predictor on a relative risk in multinomial logistic regression. It represents the ratio of the relative risks of two multinomial outcome categories between two cases with different characteristics.

Residual: A residual is the difference between a prediction based on a fitted model and an observed value. This is in contrast to an error, which is the difference between a prediction based on a true model and a true value. A residual can be considered an estimate of the error. For example, in the fitted model

$$Y_i = \hat{\beta}_0 + \hat{\beta}_1 X_i + \hat{\varepsilon}_i$$

the residuals are $\hat{\varepsilon}_i = Y_i - \left(\hat{\beta}_0 + \hat{\beta}_1 X_i\right)$. The residuals represent the part of the outcome variation that the model does not account for. As such, 'explaining the residual variation', that is, trying to explain what we can't yet explain, is one of the ways in which researchers aim to make progress in science. Residuals are also used to investigate the plausibility of model assumptions about the errors. To this end, often transformations of residuals are used, such as standardised residuals or deviance residuals.

Slope: In regression models, a slope coefficient of a predictor variable (X) tells us by how much the outcome (Y) is predicted to differ for a 1-unit difference in X, keeping all predictor variables other than X constant.

Standardisation: See *z-Standardisation*.

Transformation: See *Variable transformation*.

Variable transformation: A variable transformation is an operation that assigns new values to a variable via a mathematical function. For example, the square transformation of a variable X is calculated as trans(X) = X^2. The logarithmic transformation defines trans(X) = log (X), and so forth. In general, trans(X) = $f(X)$, where $f(\ )$ may in principle be any mathematical function. Transformations of predictors are used within regression models to enable modelling of non-linear relationships between a numeric predictor and an outcome.

z-Standardisation: z-Standardisation of a variable X means to transform X into a new variable Z by calculating $z = \frac{X - \bar{X}}{s}$, where $\bar{X}$ is the sample mean of X, and s is the sample standard deviation of X. That is, we subtract from X its mean and divide the result by the standard deviation. The mean of a z-standardised variable is equal to zero, and its standard deviation is equal to 1.

z-Test: The z-test is a type of statistical hypothesis test. The test statistic is called z and is calculated as $z = \frac{D - D_0}{SE_D}$, where D is a statistic that we wish to test, D_0 is the value of D assumed by the null hypothesis, and SE_D is the standard error of D. The test is valid under the assumption that z is normally distributed. The standard error of D needs to be known for this assumption to be valid. If SE_D is unknown and needs to be estimated from the data, then (if other assumptions are met) the test statistic follows a t-distribution, and a t-test is more appropriate than a z-test. However, the larger the sample, the closer the result of the t-test will be to the result of the z-test. The z-test is often first encountered by students of statistics as a test about the population mean, where $z = \frac{\bar{X} - \mu_0}{SE_{\bar{x}}}$, and $SE_{\bar{x}} = \frac{\sigma_x}{\sqrt{n}}$, where n is the sample size and σ_x is the population standard deviation of X. In this book, the z-test is sometimes used to test hypotheses about regression coefficients from the models introduced in Chapters 2 to 5, where $z = \frac{\hat{\beta} - \beta_0}{\widehat{SE}_{\hat{\beta}}}$. This use of the z-test relies on the assumption that the coefficients have an approximately normal sampling distribution, which is called the 'normal approximation'. This assumption is plausible when the sample is large, and when other model assumptions are met. In smaller samples, often the likelihood ratio test is a more valid test to use to investigate hypotheses about regression coefficients.

Check out the next title in the collection: *Introduction to Modern Modelling Methods*, for guidance on Conceptual Grounding and Modelling Methods.

REFERENCES

Agresti, A. (2013). *Categorical data analysis* (3rd ed.). Wiley.

Arbuthnott, J. (1710). An argument for divine providence taken from the constant regularity observ'd in the births of both sexes. *Philosophical Transactions of the Royal Society of London, 27*(328), 186–190. https://doi.org/10.1098/rstl.1710.0011

Bartlett, J. W. (2014, April 26). Deviance goodness of fit test for Poisson regression. *The Stats Geek.* https://thestatsgeek.com/2014/04/26/deviance-goodness-of-fit-test-for-poisson-regression/

Battey, H. S., Cox, D. R., & Jackson, M. V. (2019). On the linear in probability model for binary data. *Royal Society Open Science, 6*(5), Article 190067. https://doi.org/10.1098/rsos.190067

Bohannon, J. (2015, May 27). I fooled millions into thinking chocolate helps weight loss. Here's how. Gizmodo: io9. http://io9.gizmodo.com/i-fooled-millions-into-thinking-chocolate-helps-weight-1707251800

Box, G. E. P., Hunter, J. S., & Hunter, W. G. (2005). *Statistics for experimenters.* Wiley.

Bryan, J. (2017). gapminder: Data from Gapminder. R package. https://cran.r-project.org/package=gapminder

Caldwell, T. M., Rodgers, B., Clark, C., Jefferis, B. J. M. H., Stansfeld, S. A., & Power, C. (2008). Lifecourse socioeconomic predictors of midlife drinking patterns, problems and abstention: Findings from the 1958 British Birth Cohort Study. *Drug and Alcohol Dependence, 95*(3), 269–278. https://doi.org/10.1016/j.drugalcdep.2008.01.014

Carney, D. R., Cuddy, A. J. C., & Yap, A. J. (2010). Power posing. *Psychological Science, 21*(10), 1363–1368. https://doi.org/10.1177/0956797610383437

Centre for Multilevel Modelling. (n.d.). *LEMMA: Learning environment for multilevel methodology and applications.* www.cmm.bris.ac.uk/lemma/

Chao, F., Gerland, P., Cook, A. R., & Alkema, L. (2019). Systematic assessment of the sex ratio at birth for all countries and estimation of national imbalances and regional reference levels. *Proceedings of the National Academy of Sciences of the United States of America, 116*(19), 9303–9311. https://doi.org/10.1073/pnas.1812593116

Colquhoun, D. (2014). An investigation of the false discovery rate and the misinterpretation of p-values. *Royal Society Open Science, 1*(3), Article 140216. https://doi.org/10.1098/rsos.140216

Courvoisier, D. S., Combescure, S., Agoritsas, T., Gayet-Ageron, A., & Perneger, T. V. (2011). Performance of logistic regression modeling: Beyond the number of events per variable, the role of data structure. *Journal of Clinical Epidemiology, 64*(9), 993–1000. https://doi.org/10.1016/j.jclinepi.2010.11.012

Cubasch, U., Wuebbles, D., Chen, D., Facchini, M. C., Frame, D., Mahowald, N., & Winther, J.-G. (2013). Introduction. In T. F. Stocker, D. Qin, G.-K. Plattner, M. Tignor, S. K. Allen, J. Boschung, A. Nauels, Y. Xia, V. Bex & P. M. Midgley (Eds.), *Climate change 2013: The physical science basis. Contribution of Working Group I to the fifth assessment report of the Intergovernmental Panel on Climate Change* (pp. 119–158). Cambridge University Press. https://doi.org/10.1017/CBO9781107415324.007

de Graaf, N. D., de Graaf, P. M., & Kraaykamp, G. (n.d.). *Familie-Enquête Nederlandse Bevolking 1998* [Family Survey Dutch Population 1998]. https://doi.org/10.17026/dans-zzu-yw93

Diggle, P. J. (2019, June 6). *What is health data science? A biostatistician's answer* [Presentation]. Statistics and Data Science in Sickness and in Health Event, University College London, UK. www.ucl.ac.uk/population-health-sciences/sites/population-health-sciences/files/diggle_ucl_june2019_0.pdf

Efron, B., & Tibshirani, R. J. (1993). *An introduction to the bootstrap.* Chapman & Hall.

Eugene, N., Oliver, C. M., Bassett, M. G., Poulton, T. E., Kuryba, A., Johnston, C., Anderson, I. D., Moonesinghe, S. R., Grocott, M. P., Murray, D. M., Cromwell, D. A., Walker, K., & NELA team. (2018). Development and internal validation of a novel risk adjustment model for adult patients undergoing emergency laparotomy surgery: The National Emergency Laparotomy Audit risk model. *British Journal of Anaesthesia, 121*(4), 739–748. https://doi.org/10.1016/j.bja.2018.06.026

Friendly, M., Dray, S., Wickham, H., Hanley, J., Murphy, D., & Li, P. (2020). *HistData: Data sets from the history of statistics and data visualization.* https://cran.r-project.org/package=HistData

Ganzeboom, H. B. G., de Graaf, P. M., & Treiman, D. J. (1992). A standard international socio-economic index of occupational status. *Social Science Research, 21*(1), 1–56. https://doi.org/10.1016/0049-089X(92)90017-B

Gelman, A., & Hennig, C. (2017). Beyond subjective and objective in statistics. *Journal of the Royal Statistical Society. Series A: Statistics in Society, 180*(4), 967–1033. https://doi.org/10.1111/rssa.12276

Gelman, A., & Loken, E. (2013). The garden of forking paths: Why multiple comparisons can be a problem, even when there is no 'fishing expedition' or 'p-hacking' and the research hypothesis was posited ahead of time. www.stat. columbia.edu/~gelman/research/unpublished/p_hacking.pdf

Gelman, A., & Loken, E. (2014). The statistical crisis in science. *American Scientist, 102*(6), 460–465. https://doi.org/10.1511/2014.111.460

Gigerenzer, G. (2004). Mindless statistics. *Journal of Socio-Economics, 33*(5), 587–606. https://doi.org/10.1016/j.socec.2004.09.033

Gigerenzer, G., & Marewski, J. N. (2014). Surrogate science: The idol of a universal method for scientific inference. *Journal of Management, 41*(2), 421–440. https:// doi.org/10.1177/0149206314547522

Gigerenzer, G., Swijtink, Z., Porter, T., Daston, L., Beatty, J., & Krüger, L. (1989). *The empire of chance. How probability changed science and everyday life.* Cambridge University Press. https://doi.org/10.1017/CBO9780511720482

GISTEMP Team. (2020). *GISS Surface Temperature Analysis (GISTEMP)* (Version 4). NASA Goddard Institute for Space Studies. Retrieved 28 June, 2020, from https:// data.giss.nasa.gov/gistemp/

Greenland, S., Senn, S. J., Rothman, K. J., Carlin, J. B., Poole, C., Goodman, S. N., & Altman, D. G. (2016). Statistical tests, P values, confidence intervals, and power: A guide to misinterpretations. *European Journal of Epidemiology, 31*(4), 337–350. https://doi.org/10.1007/s10654-016-0149-3

Hilbe, J. M. (2009). *Logistic regression models.* Chapman & Hall. https://doi. org/10.1201/9781420075779

Hilbe, J. M. (2012). *Negative binomial regression* (2nd ed.). Cambridge University Press.

Hilbe, J. M. (2014). *Modeling count data.* Cambridge University Press. https://doi. org/10.1017/CBO9781139236065

Holland, A. C. (2014). Replication data for: The distributive politics of enforcement (Version V2, UNF:5:Z9FYtN/LeCktprosg6acVQ==) [Data set]. Harvard Dataverse. https://doi.org/10.7910/DVN/24859

Holland, A. C. (2015). The distributive politics of enforcement. *American Journal of Political Science, 59*(2), 357–371. https://doi.org/10.1111/ajps.12125

Hosmer, D. W., Lemeshow, S., & Sturdivant, R. X. (2013). *Applied logistic regression* (3rd ed.). Wiley. https://doi.org/10.1002/9781118548387

Ioannidis, J. P. A. (2005). Why most published research findings are false. *PLoS Medicine*, *2*(8), Article e124. https://doi.org/10.1371/journal.pmed.0020124

Ioannidis, J. P. A. (2019). The importance of predefined rules and prespecified statistical analyses. Do not abandon significance. *JAMA Journal of the American Medical Association*, *321*(21), 2067–2068. https://doi.org/10.1001/jama.2019.4582

Jackman, S. (2017). *pscl: Classes and methods for R developed in the Political Science Computational Laboratory*. United States Studies Centre, University of Sydney. https://github.com/atahk/pscl/

King, G. (1988). Statistical models for political science event counts: Bias in conventional procedures and evidence for the exponential Poisson regression model. *American Journal of Political Science*, *32*(3), 838–863. https://doi.org/10.2307/2111248

Kraaykamp, G., van Eijc, K., & Ultee, W. (2010). Status, class and culture in the Netherlands. In T. W. Chan (Ed.), *Social status and cultural consumption* (pp. 169–203). Cambridge University Press. https://doi.org/10.1017/CBO9780511712036.007

Krzanowski, W. J. (1998). *An introduction to statistical modelling*. Wiley.

Kuhn, M., & Johnson, K. (2013). *Applied predictive modeling*. Springer. https://doi.org/10.1007/978-1-4614-6849-3

Lamb, E. (2012, July 17). 5 sigma what's that? *Scientific American*. https://blogs.scientificamerican.com/observations/five-sigmawhats-that/

Lee, A. H., Wang, K., Yau, K. K. W., & Somerford, P. J. (2003). Truncated negative binomial mixed regression modelling of ischaemic stroke hospitalizations. *Statistics in Medicine*, *22*(7), 1129–1139. https://doi.org/10.1002/sim.1419

Lenssen, N. G., Schmidt, J., Hansen, M., Menne, A., Persin, R., Ruedy, R., & Zyss, D. (2019). Improvements in the GISTEMP uncertainty model. *Journal of Geophysical Research: Atmospheres*, *124*(12), 6307–6326. https://doi.org/10.1029/2018JD029522

Liao, T. F. (1994). *Interpreting probability models: Logit, probit, and other generalized linear models*. Sage.

Lim, C., & Putnam, R. D. (2010). Religion, social networks, and life satisfaction. *American Sociological Review*, *75*(6), 914–934. https://doi.org/10.1177/0003122410386686

Long, J. S., & Freese, J. (2014). *Regression models for categorical dependent variables using Stata* (3rd ed.). Stata Press.

MacDougal, D. W. (2012). Galileo's great discovery: How things fall. In *Newton's gravity: An introductory guide* (pp. 17–36). Springer. https://doi.org/10.1007/978-1-4614-5444-1_2

Mayo, D. G. (2018). *Statistical inference as severe testing: How to get beyond the statistics wars*. Cambridge University Press. https://doi.org/10.1017/9781107286184

McShane, B. B., Gal, D., Gelman, A., Robert, C., & Tackett, J. L. (2019). Abandon statistical significance. *The American Statistician, 73*(Suppl. 1), 235–245. https://doi.org/10.1080/00031305.2018.1527253

Nicholls, D., Statham, R., Costa, S., Micali, N., & Viner, R. M. (2016). Childhood risk factors for lifetime bulimic or compulsive eating by age 30 years in a British national birth cohort. *Appetite, 105*, 266–273. https://doi.org/10.1016/j.appet.2016.05.036

Paul, P., Pennell, M. L., & Lemeshow, S. (2013). Standardizing the power of the Hosmer–Lemeshow goodness of fit test in large data sets. *Statistics in Medicine, 32*(1), 67–80. https://doi.org/10.1002/sim.5525

Pawitan, Y. (2013). *In all likelihood. Statistical modelling and inference using likelihood.* Oxford University Press.

Rabe-Hesketh, S., & Skrondal, A. (2012). *Multilevel and longitudinal modeling using Stata*: Vol. II. *Categorical responses, counts, and survival* (3rd ed.). Stata Press.

Ranehill, E., Dreber, A., Johannesson, M., Leiberg, S., Sul, S., & Weber, R. A. (2015). Assessing the robustness of power posing. *Psychological Science, 26*(5), 653–656. https://doi.org/10.1177/0956797614553946

Royston, P., & Sauerbrei, W. (2005). Building multivariable regression models with continuous covariates in clinical epidemiology: With an emphasis on fractional polynomials. *Methods of Information in Medicine, 44*(4), 561–571. https://doi.org/10.1055/s-0038-1634008

Salih, T., Martin, P., Poulton, T., Oliver, C. M., Basset, M. G., Moonesinghe, S. R., & NELA Project Team. (2020). Distance travelled to hospital for emergency laparotomy and the effect of travel time on mortality: Cohort study. *BMJ Quality & Safety*. Advance Online publication. https://doi.org/10.1136/bmjqs-2019-010747

Sauerbrei, W., & Royston, P. (1999). Building multivariable prognostic and diagnostic models: Transformation of the predictors by using fractional polynomials. *Journal of the Royal Statistical Society. Series A: Statistics in Society, 162*(1), 71–94. https://doi.org/10.1111/1467-985X.00122

Shickle, D., & Farragher, T. M. (2015). Geographical inequalities in uptake of NHS-funded eye examinations: Small area analysis of Leeds, UK. *Journal of Public Health (Oxford, England), 37*(2), 337–345. https://doi.org/10.1093/pubmed/fdu039

Silver, N. (2012). *The signal and the noise: The art and science of prediction.* Penguin.

Simmons, J. P., Nelson, L. D., & Simonsohn, U. (2011). False-positive psychology: Undisclosed flexibility in data collection and analysis allows presenting anything as significant. *Psychological Science, 22*(11), 1359–1366. https://doi.org/10.1177/0956797611417632

Stigler, S. M. (2016). *The seven pillars of statistical wisdom.* Harvard University Press. https://doi.org/10.4159/9780674970199

Tabachnick, B. G., & Fidell, L. S. (2014). *Using multivariate statistics* (6th ed.). Pearson.

UK Department of Communities and Local Government. (2011). *The English Indices of Deprivation 2010. Neighbourhoods statistical release.* www.gov.uk/government/ statistics/english-indices-of-deprivation-2010

Van Smeden, M., De Groot, J. A. H., Moons, K. G. M., Collins, G. S., Altman, D. G., Eijkemans, M. J., & Reitsma, J. B. (2016). No rationale for 1 variable per 10 events criterion for binary logistic regression analysis. *BMC Medical Research Methodology, 16*(1), Article 163. https://doi.org/10.1186/s12874-016-0267-3

Venables, W. N., & Ripley, B. D. (2002). *Modern applied statistics with S* (4th ed.). Springer. https://doi.org/10.1007/978-0-387-21706-2

von Bortkiewicz, L. (1898). *Das Gesetz der kleinen Zahlen.* B. G. Teubner.

Wasserstein, R. L., & Lazar, N. A. (2016). The ASA's statement on p-values: Context, process, and purpose. *The American Statistician, 70*(2), 129–133. https://doi.org/10 .1080/00031305.2016.1154108

Williams, R. (2006). Generalized ordered logit/partial proportional odds models for ordinal dependent variables. *Stata Journal, 6*(1), 58–82. https://doi.org/10.1177/1 536867X0600600104

Williams, R. (2016). Understanding and interpreting generalized ordered logit models. *Journal of Mathematical Sociology, 40*(1), 7–20. https://doi.org/10.1080/00 22250X.2015.1112384

Wilson, P. (2015). The misuse of the Vuong test for non-nested models to test for zero-inflation. *Economics Letters, 127,* 51–53. https://doi.org/10.1016/j. econlet.2014.12.029

Wood, S. N. (2006). *Generalized additive models: An introduction with R.* Chapman & Hall.

Zaninotto, P., & Falaschetti, E. (2011). Comparison of methods for modelling a count outcome with excess zeros: Application to Activities of Daily Living (ADL-s). *Journal of Epidemiology & Community Health, 65*(3), 205–210. https://doi.org/10.1136/ jech.2008.079640

INDEX

Page numbers in *italic* indicate figures and in **bold** indicate tables.

CPSIA information can be obtained
at www.ICGtesting.com
Printed in the USA
JSHW062310130723
44476JS00001B/39

9 781529 761269